Organic Evolution

Prof. B.L. Chaudhary, MSc, PhD, F.B.S.

Chairman, Board of Secondary Education Ajmer
and
Ex-Vice Chancellor, ML Sukhadia University
Udaipur (Raj).

Dr. Kailash Choudhary

Director, Life Sciences
Institute for Advanced Studies (I*f*AS)
Pune (Mah.)

Dr. Arun Chaudhary

Assistant Professor
Department of Botany
Government College, Chittorgarh (Raj.)
(CSIR-NET & GATE- Life Sciences)

SCIENTIFIC PUBLISHERS

Published by

SCIENTIFIC PUBLISHERS (INDIA)
5 A, New Pali Road
P.O. Box 91
Jodhpur - 342 001 INDIA

ISBN:9789387893870
9 789387 893870
Price: 1650/-

Visit the Scientific Publishers (India) website at
http://www.scientificpub.com
Printed in India

PREFACE

Organic evolution is an exciting and interesting stream of biology; from birth of human civilization, philosophers, scientists, and naturalists starts thinking about the form, diversity and origin of life and put forward many theories to explain origin and evolution of life forms. The theory of evolution is outstandingly the most important theory in biology, which helps in solving puzzles of origin of life, evolution of populations and speciation.

Many books are available on Organic Evolution both by Indian and Foreign authors for students with different backgrounds at the undergraduate and post graduate levels, however, these books cover one or few specialized areas rather than covering the broad spectrum of evolution. Consequently, the students have to depend on many books for updated information on evolution, so there is a long-felt need for a comprehensive book on evolution, with the latest information. Therefore keeping in mind the demand of updated information, we bring this comprehensive book for students.

The present book "Organic evolution", with several novel features, is comprehensively written with latest advances in the subject and is divided in seven chapters. Chapter 1 of the book describes the environment of primitive earth and its importance for origin of first life form. Chapter 2 includes evolutionary theories and supporting evidences in favour of evolution. The role of population genetics in evolution of gene pools of population is the subject of Chapter 3. The concept of species and modes of speciation, origin of new species and required changes in genetic for speciation is discussed in Chapter 4. Chapter 5 provides the essential information needed to understand molecular evolution, and molecules as a reliable tool for molecular systematic and in reconstruction of phylogenies. Chapter 6 of the book describes vertebrate diversity and vertebrate evolution, primarily focusing on Primates and the evolution of *Homo sapiens*. Finally, Chapter 7 is about origin and radiation of Angiosperms. The language and contents of this book are made so simple and easy that even a lay man with a minimal knowledge of Life Sciences will be able to understand the essence of evolution.

This book has been designed mainly to provide the most fundamental and updated knowledge of the subject to undergraduate and postgraduate students in various streams of Life Sciences such as Botany, Zoology, Biotechnology, Microbiology, Biochemistry, Agriculture etc, of many Indian Universities. Besides this book will be beneficial to student's preparing for CSIR NET Life

Sciences, GATE Life Sciences, ICMR Life Sciences, IAS, IFS and other State Civil service examination.

Authors are thankful to Shri Pawan Kumar Sharma and Tanay Sharma of Scientific Publisher (India), Jodhpur for timely bringing out of this book in nice getup.

Authors invite the readers to understand and learn the very basics of evolution to maximum possible with ease and pleasure. As this book is intended to cater the needs of students learning evolution from many different disciplines, there is a lot of scope to improve the book in future.

We welcome useful suggestions and healthy criticism to improve the book.

Prof. B.L. Chaudhary
Dr. Kailash Choudhary
Dr. Arun Chaudhary

CONTENTS

1.	**Early earth and the origin of life**	**1-19**
	1.1 Major events in the history of life: a preview	3
	1.2 The origin of life	7
	1.2.1 Theory of special creation	7
	1.2.2 Theory of spontaneous generation	7
	1.2.3 Theory of biogenesis	7
	1.3 Primitive environment and chemical evolution	7
	1.3.1 Modern theory or Naturalistic theory of origin of life	8
	1.3.2 Formation of Earth	8
	1.3.3 Synthesis of simple organic compounds	10
	1.3.4 Synthesis of organic monomers	11
	1.3.5 Synthesis of organic polymers	12
	1.3.6 Aggregation of organic polymers	13
	1.4 RNA may have been the first genetic material	14
	1.5 Controversy about the origin of life	16
2.	**Evolutionary theories and evidences in favour of evolution**	**20-50**
	2.1 Lamarckism	21
	2.1.1 Main postulates of Lamarckism	22
	2.1.2 Criticism	22
	2.2 Darwinism	23
	2.2.1 Main postulates of Darwinism	25
	2.2.2 Artificial selection	26
	2.2.3 Sexual selection	26
	2.2.4 Critics of Darwinism	27
	2.3 The Modern synthesis	28
	2.4 Evidences in favour of evolution	29
	2.4.1 On small scale: Natural selection provides evidence of evolution	30

2.4.1.1. The evolution of insecticide-resistant insects 30
2.4.1.2 The evolution of drug-resistant HIV 30
2.4.2 Evidences of large scale evolutionary changes 31
2.4.2.1 Evidences from Biogeography 31
2.4.2.2 Evidences from Fossils 37
2.4.2.3 Geological time scale 43
2.4.2.4 Evidences from comparative anatomy 45
2.4.2.5 Evidences from embryology 48
2.4.2.6 Evidences from Biochemistry and molecular biology 49

3. The Mechanism of evolution 51-95

3.1 The foundation of genetics 52
3.1.1 Mendel's Experiments and the laws of inheritence 53
3.1.2 Monohybrid cross 54
3.1.3 Dihybrid cross 56
3.1.4 Some Exception to Mendal's Laws 59
3.2 Population genetics 63
3.2.1 Integration of Darwinian selection and Mendelian inheritance to give rise modern evolutionary theory 63
3.2.2 Genetic structure of a population 64
3.2.3 Hardy-Weinberg theorem for non evolving populations 65
3.2.4 Hardy-Weinberg equilibrium for two loci 67
3.2.5 Hardy-Weinberg equilibrium for multi loci 68
3.2.6 Conditions required to maintain Hardy Weinberg equilibrium 68
3.3 Microevolution and its causes 70
3.3.1 Genetic drift 71
3.3.2 Gene flow 74
3.3.3 Mutation 74
3.3.4 Non random mating 75
3.3.5 Natural selection 76
3.4 Macroevolution 77
3.4.1 Mammal's evolution from reptilian ancestors: a closer look 78
3.5 Extinction and Radiations 80
3.5.1 Cambrian explosion: One of the most important adaptive radiations in the history of life. 82
3.5.2 Mass extinctions 83
3.5.2.1 Permian extinction 84

3.5.2.2 Extinction of dinosaurs (The Cretaceous extinction) 85
3.5.2.3 Sixth mass extinction 86
3.6 Coevolution 86
3.6.1 Insect-plant coevolution 87
3.7 Adaptation and Natural selection 88
3.7.1 Adaptation 88
3.7.2 Types of adaptation 89
3.7.3 Coadaptation 90
3.8 Natural selection as the mechanism of adaptive evolution 91
3.8.1 Concept of fitness 93
3.8.2 The effect of selection on a varying characteristic can be stabilizing, directional, or diversifying 93

4. The Origin of species 96-121

4.1 Species concept 97
4.1.1The biological species concept 97
4.1.2 Recognition species concept 99
4.1.3 The phenetic species concept 99
4.1.4 The ecological species concept 100
4.2 Reproductive isolation barriers 101
4.2.1 Prezygotic Barriers 101
4.2.2 Postzygotic Barriers 103
4.3 Modes of speciation 104
4.3.1 Geographical barriers leading to Allopatric speciation 105
4.3.2 Adaptive radiation on island chains 107
4.3.3 Parapatric speciation 109
4.3.4 Sympatric speciation 111
4.4 Factors facilitating speciation 114
4.5 How much genetic change is required for speciation? 116
4.5.1 Punctuated equilibrium model 116
4.6 Polymophism 119
4.6.1 Balanced polymorphism 121

5. Molecular evolution and Phylogenies 122-154

5.1 Molecular Evolution 123
5.1.1 Neutral Theory of Molecular Evolution 123
5.1.2 The Evolution of Macromolecules 125

5.1.3 Molecular evolution is driven by changes in nucleotide 126

5.1.4 Many mutations may be selectively neutral 127

5.1.5 Determining and Comparing the Structure of Macromolecules 128

5.2 Changes in macromolecules can serve as molecular clocks 131

5.3 Molecular Systematics and phylogenies 132

5.3.1 Homologous traits are inherited from a common ancestor 132

5.3.2 Identifying ancestral traits is sometimes difficult 134

5.4 Steps in Reconstructing Phylogenies 135

5.4.1 Morphological and developmental traits are used in reconstructing phylogenies 135

5.4.2 Molecular traits are also useful in reconstructing phylogenies 136

5.4.3 Reconstructing a Simple Phylogeny 137

5.4.4 Systematists use the parsimony principle when reconstructing phylogenies 139

5.5 Biological Classification and Evolutionary Relationships 140

5.5.1 Current biological classifications reflect evolutionary relationships 141

5.6 Phylogenetic Trees Have Many Uses 143

5.6.1 How many times has a trait evolved? 144

5.6.2 When did lineages split? 144

5.7 Genomes and their Evolution 145

5.7.1 Proteins Acquire New Functions 145

5.7.2 The Evolution of Genome Size 148

5.7.3 Gene duplication can increase genome size and complexity 150

5.8 The Uses of Molecular Genomic Information 151

5.8.1 Molecular information is used to reconstruct phylogenies 152

5.8.2 Molecular data are used to determine the phylogenetic histories of genes 152

5.8.3 Molecular information provides new ways to combat diseases 153

6. Vertebrate evolution and diversity 155-199

6.1 Invertebrate Chordates and the origin of vertebrates 155

6.1.1 Four anatomical features characterize phylum Chordata 156

6.1.2 Invertebrate chordates provides clues to the origin of vertebrates 157

6.2 Introduction to the Vertebrates 160

6.2.1 Characteristics of Subphylum Vertebrates 160

6.2.2 Overview of vertebrate diversity 162

6.3 Superclass Agnatha: Jawless Vertebrates 162

6.3.1 Lampreys and hagfish are the only extant agnathans 163

6.4 Superclass Gnathostomata I: The Fishes 163

6.4.1 Evolution of vertebrate jaws from skeleton support of the pharyngeal slits 164

6.4.2 Class Chondrichthyes: A cartilaginous endoskeleton 165

6.4.3 Class Osteichthyes: A bony endoskeleton 167

6.5 Superclass Gnathostomata II: The Tetrapods 169

6.5.1 Amphibians are the oldest class of tetrapods 169

6.5.2 Evolution of the amniotic egg expanded the success of vertebrates on land. 172

6.5.3 Reptilian evolution 173

6.5.4 Evolution of Birds 177

6.5.5 Mammalian evolution 180

6.6 Primates and the evolution of *Homo sapiens* 184

6.6.1 Primate evolution provides a context for understanding human origins 184

6.6.2 Humanity: a very young twig on the vertebrate tree 187

6.6.3 Four main classes of change occurred during hominin evolution 190

6.6.4 Occurrence of hominids to be classified as genus *Homo* 193

6.6.5 The Emergence of anatomically modern humans: *Homo sapiens* 195

6.7 Cultural Evolution 199

7. Origin and Evolution of Angiosperms 201-222

7.1 Introduction: An Overview of evolution of land plants 201

7.2 Fossil evidences of Angiosperms 203

7.2.1 The Koonwara Angiosperm 204

7.2.2 Features of other fossil flowers 205

7.3 The origin and radiation of Angiosperms 207

7.3.1 Time period of Angiosperms evolution 207

7.3.2 Probable site of Angiosperms evolution 208

7.3.3 Causes of Angiosperms evolution and diversification 209

7.3.4 Rise of Angiosperms to Ecological dominance 210

7.4 Evolutionary transformation of a Gymnosperm into a flowering plant 211

7.5 Angiosperm Phylogeny 216

7.5.1 The basal angiosperms 217

Subject Index 223-229

Chapter 1

EARLY EARTH AND THE ORIGIN OF LIFE

1.1 Major events in the history of life: A preview

1.2 The origin of life

  1.2.1 Theory of special creation

  1.2.2 Theory of spontaneous generation

  1.2.3 Theory of biogenesis

1.3 Primitive environment and chemical evolution

  1.3.1 Modern theory or Naturalistic theory of origin of life

  1.3.2 Formation of Earth

  1.3.3 Synthesis of simple organic compounds

  1.3.4 Synthesis of organic monomers

  1.3.5 Synthesis of organic polymers

  1.3.6 Aggregation of organic polymers

1.4 RNA may have been the first genetic material

1.5 Controversy about the origin of life

INTRODUCTION

Life originated very early in earth's history, and those first organisms were ancestors of our present day organisms. The organisms most familiar to us are macroscopic and multicellular, mainly plants and animals. However, for the first three-quarters of evolutionary history, earth's only organisms were microscopic and unicellular. Earth was formed around 4.5 billion years ago, and life probably began only a few hundred million years later. There is no evidence of life for 1 billion years after the origin of earth. This is explained by the fact that after the origin of Earth, rocks present in earth get melted and after cooling go arranged according to their density, a process which is called differentiation. Scientists in 3.8 billion years old rocks in Greenland have found isotopes of carbon that seems to represent the metabolic activity of organisms. One would guess from the relatively simple structure of the prokaryotic cell that the earliest organisms were prokaryotes, and fossil records also support this presumption. Evidence of ancient prokaryotic life has been found in rocks called Stromatolites. Stromatolites are banded domes of sedimentary rocks that are similar to layered mats formed today in salt marshes and some warm ocean lagoons by colonies of bacteria and cyanobacteria. The layers are sediments that stick to the jellycoats of motile microbes, which continually migrate out of one layer of sediment and form a new one above, producing the banded pattern (Figure 1.1). Fossils resembling spherical and filamentous prokaryotes have been found in 3.5 billion years old stromatolites in southern Africa and western Australia (Figure 1.2). These are currently the oldest known fossils of living organisms. On the basis of these fossil evidences it is considered that life on earth originated between 3.5 and 4 billion years ago.

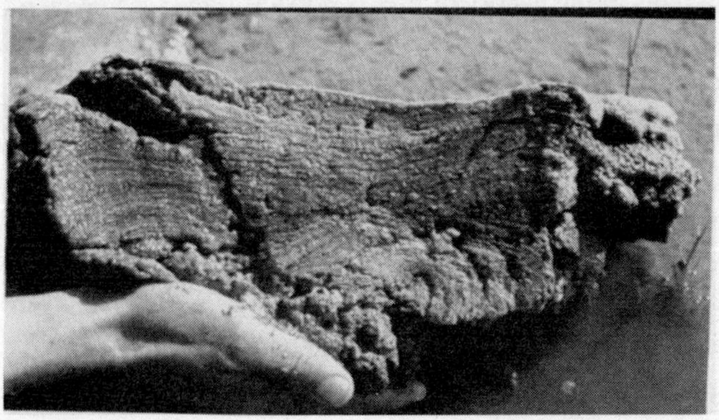

Figure 1.1. Fossilized mats known as stromatolites. The bands seen in this section of a mat are alternating layers of sediment that adhere to the sticky prokaryotes, which produce the succession of layers by migrating upwards.

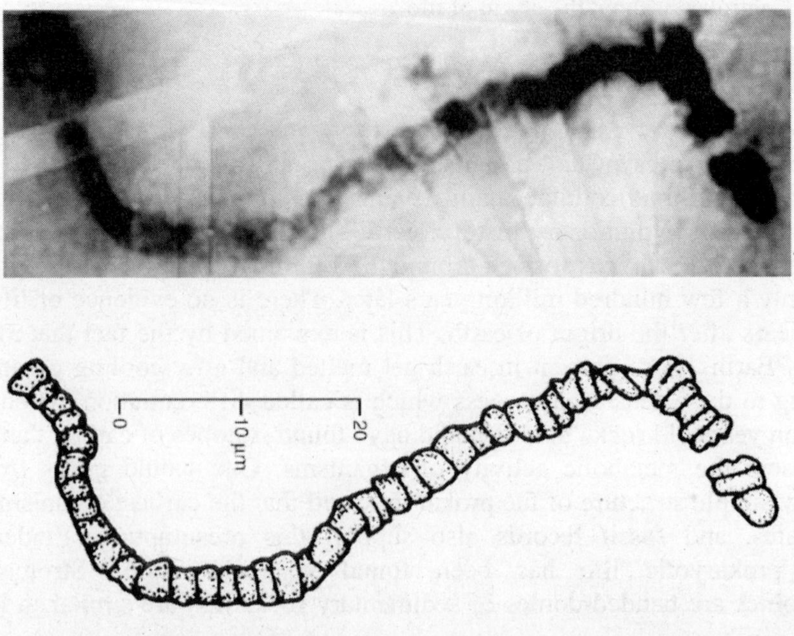

Figure 1.2. This fossilized filamentous prokaryote, about 3.5 billion years old, was found in a stromatolite collected in western Australia

1.1 MAJOR EVENTS IN THE HISTORY OF LIFE: A PREVIEW

Before we focus on the question of how life on earth began, let's preview some major episodes of biological history in the context of geological time (Figure 1.3).

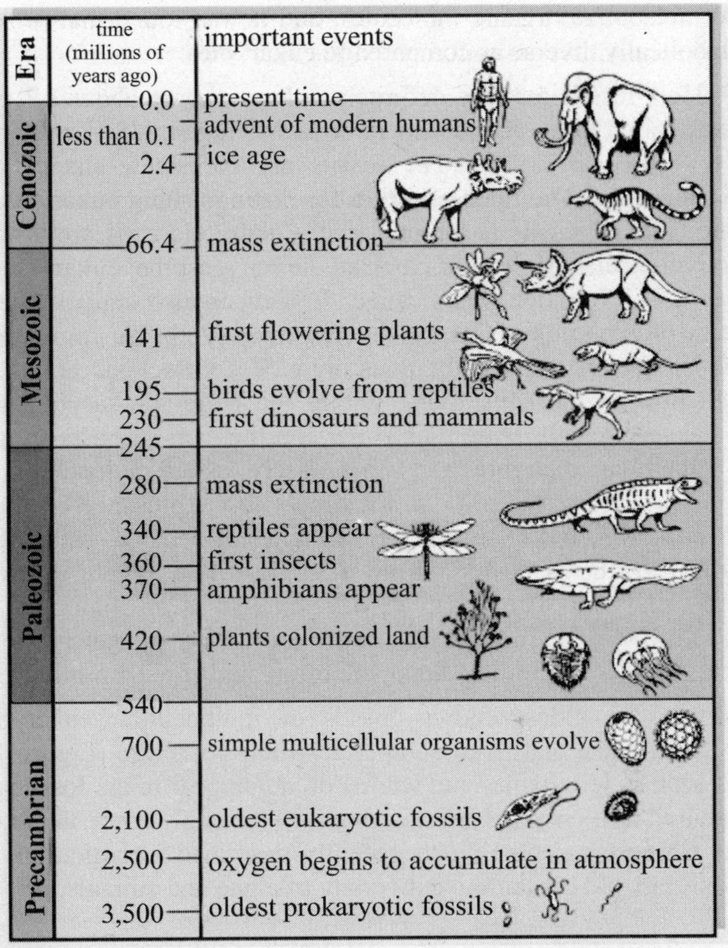

Figure 1.3. Some major episodes in the history of life. The timings of some events are based on fossil evidences and some on molecular clocks.

Fossil evidences as well as laboratory researches on the kind of chemical reactions that may have taken place on early earth to give rise to molecular building blocks of life (such as amino acids, sugars and nucleotides) suggest that the first life form which was prokaryotes originated within a few hundred million years after earth's crust cooled and solidified, beginning a succession of microorganisms that were the only forms of life on earth for about 2 billion years. RNA sequence data show that two distinct groups of prokaryotes "Bacteria and Archae" diverged early in the history of life. On the basis of fossil records and molecular data this split occurred between 3.0 to 2.0 billion years ago.

Oxygen revolution which was the production of free oxygen in atmosphere by early photosynthetic prokaryotes took place about 2.5 billion years ago. It was a turning point in the history of life and it laid the platform of evolution of aerobic life. As prokaryotic evolution continued, some organisms were able to tolerate the corrosive effects of atmospheric oxygen and became the first to use oxygen to metabolize organic molecules, and it was found that prokaryotes are more metabolically diverse as compared to eukaryotes.

The oldest fossils that are definitely eukaryotic are about 1.7 billion years old and was found in an abandoned mine in Michigan, USA, and are described by Han & Runnegar (1992). The fossils are corkscrew shaped and closely resemble later algae. The main criterion for distinguishing eukaryotic cells from prokaryotic cells in fossils is the cell size. Eukaryotic cells are typically larger than prokaryotic cells. Molecular clock studies suggest that eukaryotes originated in between 2.2–1.7 billion years range. Evidences also support the Endosymbiotic theory of evolution of eukaryotic cells from symbiotic community of prokaryotes which created the membranous organelles mitochondria and chloroplast from small prokaryotic cells living within the larger prokaryotic cells. These primitive eukaryotic cells were unicellular and the name protists are given to the group of unicellular eukaryotes and some closely related multicellular organisms. Multicellular life is used to refer to not simply to the presence of more than one cell in an organism, but to more than one kind of cell, i.e. cell differentiation. Multicellular organisms develop from a single celled zygote to an adult with specialized cell types. The origin of development is an important step in the evolution of life. All life forms with cell differentiation are eukaryotic. The molecular clock suggests that multicellular life originated about 1.5 billion years ago.

However, the oldest known fossils of multicellular eukaryotes are of relatively small algae that lived about 1.2 billion years ago (Figure 1.4). Larger organisms such as jelly fishes and worms do not appear in the fossil records until several hundred million years later, in the late Precambrian era, about 600 million years ago. Chinese paleontologists recently described a particularly rich fossil 570 million years old containing a diversity of algae and animals.

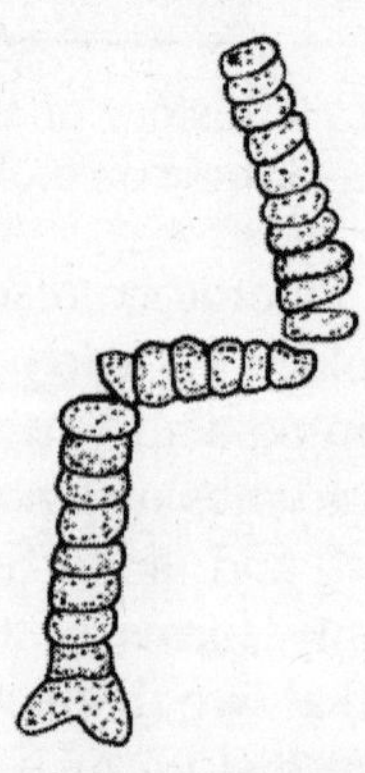

Figure 1.4. Fossilized alga about 1.2 billion years old.

A **Snowball Earth** hypothesis was proposed which explains that glaciers covered the planets landmasses from pole to pole. This hypothesis was supported by geologists who have recently reported evidences of a severe ice age 750 to 570 million years ago and it helps to explain why multicellular eukaryotes were limited in diversity and distribution until the very late Precambrian period, during this period seas probably also would have been ice covered. Thus, most life forms would have been confined to areas near deep-sea vents and hot springs. According to this hypothesis, the fossil record for the first major diversification of multicellular eukaryotes corresponds to the time of the end of Snowball Earth. A second radiation of eukaryotic forms produced most of the major groups of animals during the early Cambrian period, the first period of the Paleozoic era. However, Cnidarians and poriferans were already present in late Precmabrian period but a much greater diversity of animals evolved just after the start of the Paleozoic era 543 million years ago (Figure 1.5).

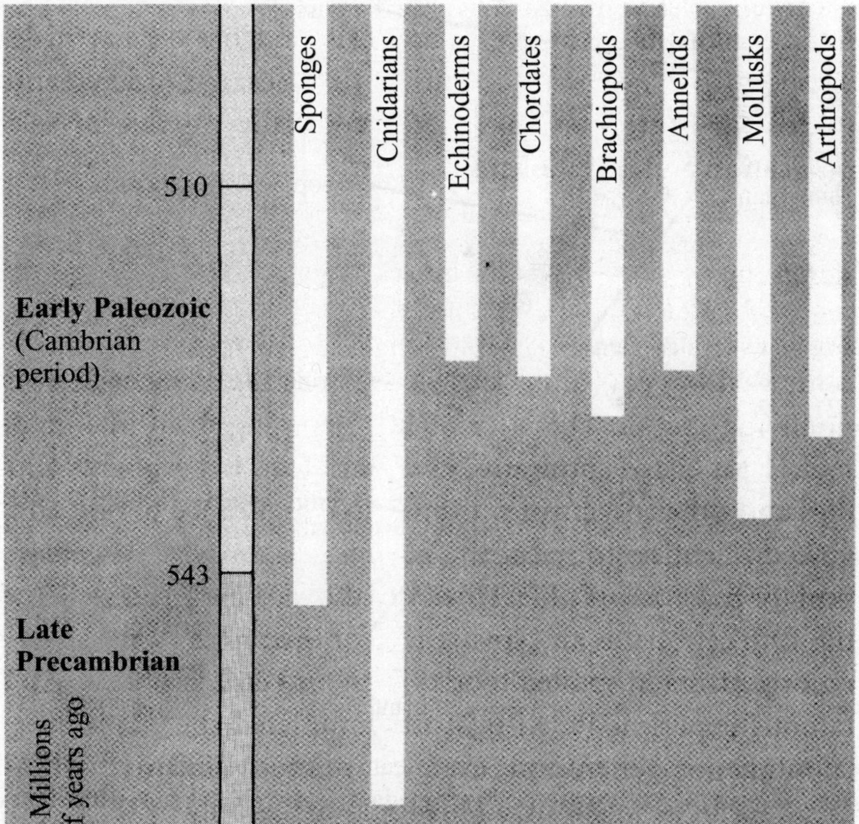

Figure 1.5. The Cambrian radiation of animals

The colonization of land was an important milestone in the history of life. Fossil evidences suggest that terrestrial life existed only in microbial forms (cynobacteria and photosynthetic prokaryotes) over a billion years ago until the land was colonized by plants and fungi. However, macroscopic life in the form of plants, fungi, and animals did not colonize land until about 500 million years ago. Plants colonized land in the company of fungi. Even today evidences are there showing association of fungi with roots of most plants. Such symbiotic association of plants and fungi are evident in some of the oldest fossilized roots. Land plants are more closely related to a group of green algae called charophyceans (Figure 1.6). Early fossil spores of 475 million years ago have a "tetrad" structure that is found in some modern bryophytes, and most early plant fossils seem to be related to the phylogenetic branches leading to modern bryophytes and pteridophytes. The main events in the early evolution of land plants were the evolution of a resistant spore stage, then the evolution of vascular tissue, followed by roots and then leaves.

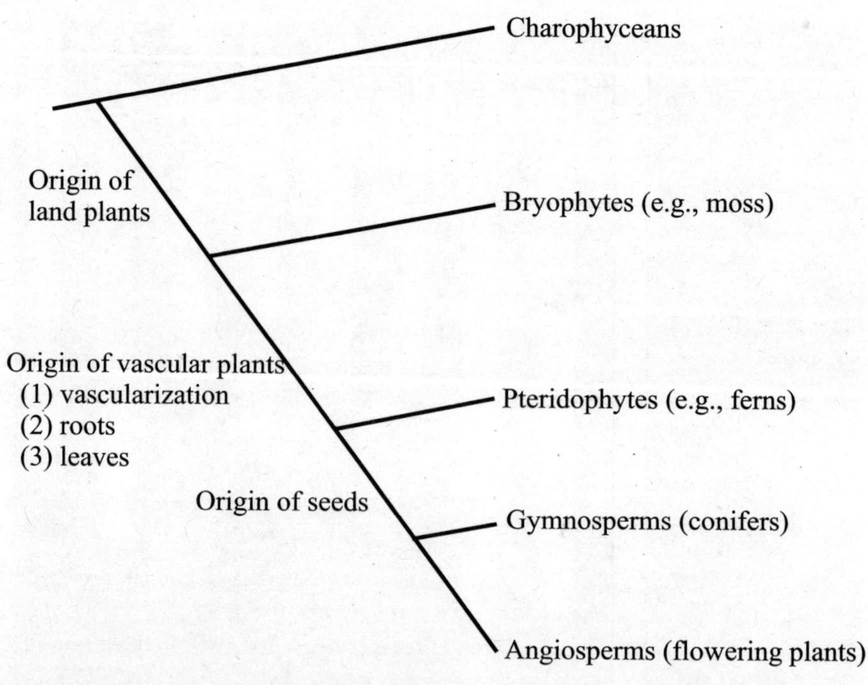

Figure 1.6. Phylogeny of the main groups of plants.

Plants transformed the landscape, creating new opportunities for all life, including herbivorous animals and their predators. Although many animal groups are represented in terrestrial environments, the most widespread and diverse land animals are certain arthropods and vertebrates. The terrestrial vertebrates are called **tetrapods**, a reference to the four walking limbs that distinguish them from the various aquatic vertebrates. An excellent fossil record documents the descent of amphibians from fishes. Reptiles evolved from amphibians and both birds and mammals evolved from reptiles.

1.2 THE ORIGIN OF LIFE (BIOPOIESIS)

The question of how life began is more specifically about the genesis of prokaryotes. Sometimes between approx. 4.0 to 3.5 billion years ago, when the planet was inhabited by bacteria advanced enough to build stromatolites, the first organisms which came into existence. What was their origin? To explain this, many theories were proposed during different times of which Naturalistic or the Chemosynthetic origin given by A.I. Oparin and J.B.S. Haldane in 1920 was widely accepted. Before looking at this theory let us discuss briefly some of the other theories.

1.2.1 Theory of Special Creation

It states that life was created by some divine decree. This theory was based on religious thoughts and was proposed by Father Saurej and supported by Milton.

1.2.2 Theory of Spontaneous Generation (Abiogenesis)

From the time of the ancient Greeks until well into the 19th century, it was common knowledge that life could arose from nonliving matter. This idea of life emerging from inanimate material is called spontaneous generation. This theory was supported by Greek philosophers like Anaximender (611-547 BC), Aristotle (384-322 BC), Thales (624-547 BC), Xenophanes, Plato (427-347 BC) etc. They believed that snails, fishes, frogs etc. arose spontaneously from mud. It was believed that not only life originated spontaneously in the past but that it is originating even today. Cases were cited of the birth of grasshoppers, and many other insects and even mice and toads from a rainy spell. In the 17th Century the Abiogenesis theory was supported by Von Helmont (1642) who claimed that mice formed in 21 days from a dirty, sweat soaked shirt in a wheat barn in the dark.

1.2.3 Theory of Biogenesis

According to the theory of biogenesis life originated from pre-existing life. This theory was supported by some scientists like Francisco Redi (1668), Lazzaro Spallanzani (1767), Louis Pasteur (1862), Harvey (1651) and Huxley (1870). Louis Pasteur's famous experiments with broth completely overturned the concept of spontaneous generation, even for microorganisms. As far as we know, all life today arises only by the reproduction of preexisting life. This life from life principle is biogenesis. But what about the first organisms? If they arose by biogenesis, then they couldn't have been the first organisms. So this put a question mark on the theory of biogenesis. To solve this puzzle A.I. Oparin and J.B.S. Haldane put forward their theory of the origin of life.

1.3 PRIMITIVE ENVIRONMENT AND CHEMICAL EVOLUTION

Most biologists support the hypothesis that life on earth developed from nonliving materials (abiogenesis) that became ordered into molecular aggregates that was eventually capable of self-replication and metabolism. As far as we know, life cannot arise by spontaneous generation from inanimate material, but conditions were very different when earth was only a billion year old. Primitive earth's atmosphere was totally different from present-day atmosphere. There was very little free atmospheric O_2, for instance, and lightning, volcanic activity, meteorite bombardment, and ultraviolet radiation were all more intense than what we experience today. In that ancient environment, origin of life was possible. It was considered that the first organisms were products of a chemical evolution in four stages:

(1) Abiotic synthesis and accumulation of small organic molecules, or monomers, such as amino acids and nucleotides.

(2) Joining of these monomers into polymers, including proteins and nucleic acids.

(3) Aggregation of abiotically produced molecules into droplets, called protobionts, that had chemical characteristic different from their surroundings.

(4) Origin of heredity and multiplication.

1.3.1 Modern Theory or Naturalistic Theory of Origin of Life

In 1920 A.I. Oparin of Russia and J.B.S. Haldane of Great Britain independently postulated that condition on the primitive earth favored chemical reactions that synthesized organic compounds from inorganic precursors present in the early atmosphere and seas. This cannot happen in the modern world, Oparin and Haldane reasoned, because the present atmosphere is rich in oxygen produced by photosynthetic life. The oxidizing atmosphere of today is not conducive to the spontaneous synthesis of complex molecules because oxygen attacks chemical bonds, extracting electrons. Before oxygen-producing photosynthesis, earth had a much less oxidizing atmosphere, derived mainly from volcanic vapors. Such a reducing (electron adding) atmosphere would have enhanced the joining of simple molecules to form more complex ones. Even with a reducing atmosphere, making organic molecules would require considerable energy, which was probably provided by lightning and the intense UV radiation that penetrated the primitive atmosphere. The modern atmosphere has a layer of ozone produced from oxygen, and this ozone shield screens out most UV radiation. There is also evidence that young Sun emit more UV radiation than the old Sun. According to Oparin and Haldane, chemical conditions and energy resources was perfect for abiotic synthesis of organic molecules. Before discussing chemical evolution let us have a look at chemical and structural changes in primitive earth which built a platform for chemical evolution on earth.

1.3.2 Formation of Earth

The earth is a planet, so it was also formed just like other planets. In the beginning, earth was a huge collection of cold objects called **planetesmials**. These planetesmials which formed the earth came from two regions – the regions occupied by the inner planet's like Mercury and Venus, and from the farther regions of Jupiter, Saturn and Uranus. The planetesmials coming from the inner planets region consisted mostly of the compounds of iron, silicon and magnesium with traces of other elements whereas those coming from the regions of outer planets largely consisted of matter like hydrogen, water, carbon dioxide and methane. As more and more planetesmials collided with the earth and got stuck into it, their kinetic energy was converted into heat energy. In addition to this, the radioactive disintegration of the elements like uranium, thorium and potassium, produced a lot of heat energy. The enormous heat energy thus produced along with gravitational compression, heated the earth so much that it melted to form a liquid. This melting of earth happened about 800 million years after its formation. It was because of the melting of earth that the surface records of the earth for the first 800 million years of its existence were completely destroyed.

When the earth melted, the molten material present in it began to reorganize itself under the influence of gravity. The molten iron, being the heaviest material, fell to the centre of the earth and formed the core of the earth (innermost part of the earth). The lighter constituents which were earlier present in the core, came up and formed the crust of earth. The medium density matter remained in between the core and crust to form the mantle of earth. This organization of molten Earth into various layers having different densities is called **differentiation**. As a result of differentiation, the earth was divided into three major layers: *crust, mantle* and *core* (Figure 1.7). During this differentiation process the lightest materials present in the original earth were expelled as gases to form the first atmosphere. The newly formed atmosphere at this stage consisted mainly of gases like nitrogen, carbon dioxide, hydrogen and water vapor. There was no oxygen gas present in the initial atmosphere of earth and this was known as **evolution of atmosphere**. The crust is made up of light rocks rich in silica and aluminium, the mantle of earth consists of rocks which are largely made up of iron silicate and magnesium silicate, the core of the earth is made up of mainly iron and some nickel. The core of the earth actually consists of two parts: the **inner core** and the **outer core**. The inner core is a solid sphere of highly compressed iron whereas the outer core is of molten iron. The temperature at the centre of the earth is about 4000°C and the pressure is extremely high, being about 3.7 million atmospheres. It is because of this extremely high pressure that the iron in the inner core can remain in solid state despite the very high temperature of 4000°C (which is much above the melting point of iron, 1539°C). Since the pressure is comparatively lower in the outer core, therefore, the iron metal remains in the molten state. Thus the outer core behaves like a liquid. Since the core of the earth is very hot but the surface of earth is colder, so heat flows outwards from the hot interior of the earth to the surface of the earth. This flow of heat generates convectional currents in the interior of earth. These convectional currents are responsible for drifting of plates in the crust region (Continental drift), and thus events like volcanoes, earthquakes, formation of mountains and plateaus occur.

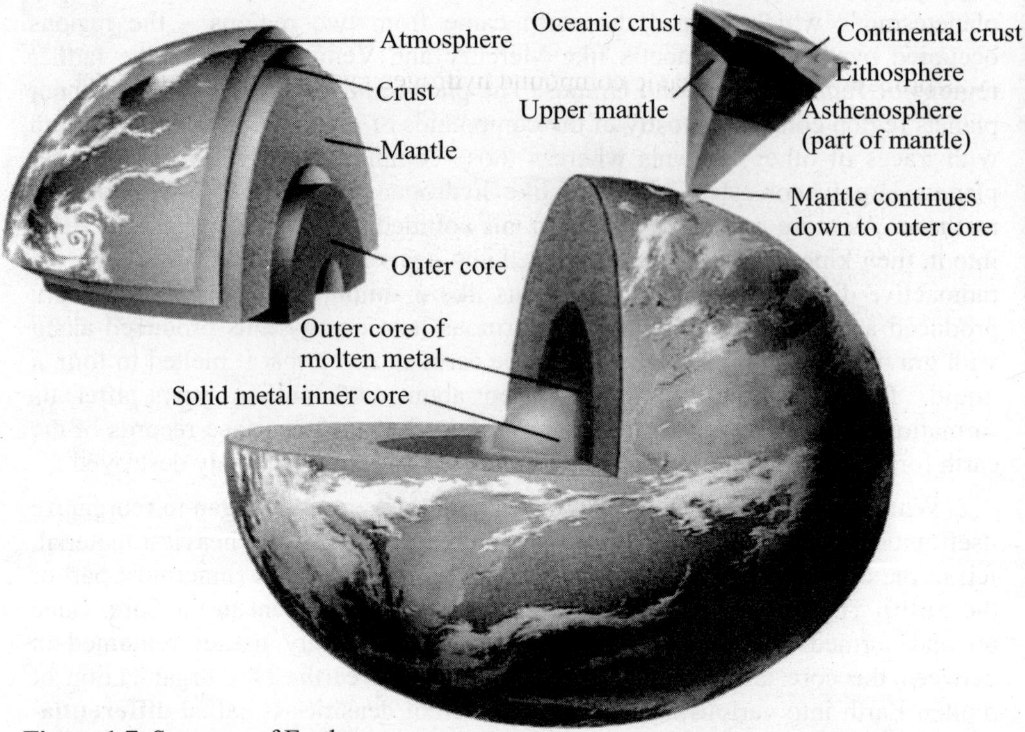

Figure 1.7. Structure of Earth

After the evolution of atmosphere it was the turn of evolution of water bodies (Oceans). The primitive crust of earth was constantly punctured by meteorites falling from the outer space. Now, since the atmosphere was very thin at that time, it could not slow down the falling meteorites by offering sufficient air resistance. Due to this, the meteorites fell on earth with great force and made many holes in the newly formed crust. These holes in the earths surface were called vents. Through these vents, the molten rocks from upper core and lower mantle, came above the surface of the earth. This molten rock called magma cooled down, solidified and became a part of the crust. This process of the magma rising to the surface of the earth through vents, released more gases to the atmosphere. These gases were mainly nitrogen, carbon dioxide and water vapors. In due course of time, a lot of water vapor was concentrated in the atmosphere. When this water vapor present in the atmosphere rose up, it got cooled. The cooled water vapor turned into clouds and then poured on the earth in the form of rain. This rain water collected in all low-lying areas of the earth's crust formed oceans. Formation of oceans was the key point in origin of first life forms because life originate in sea water.

1.3.3 Synthesis of simple organic compounds

Initially gases like hydrogen, carbon dioxide and nitrogen were abundant. These gases interacted with each other to form methane, ammonia, hydrogen cyanide, nitrides, carbides, and various other gases in dissolved form in water. Methane was probably the first organic compound hydrogen cyanide was formed later.

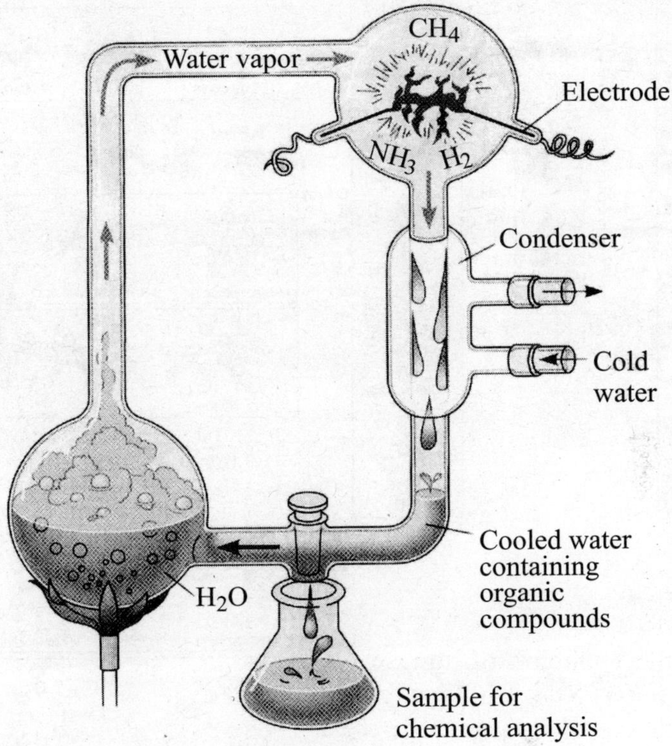

Figure 1.8. Miller and Urey experimental apparatus.

1.3.4 Synthesis of organic monomers

High concentration of these organic compounds in sea water induced them to react with each other and produce some unsaturated hydrocarbons like ethylene, aldehydes, ketones, alcohols and organic acids. As a result of further interaction of these saturated and unsaturated compounds, organic monomers like simple sugars (glucose, ribose), amino acids, fatty acids, glycerol and nitrogenous bases were synthesized. In 1953, the first support to Oparin and Haldane's theory came from Stanley Miller and Harold Urey. They created primitive earth's environmental condition in a laboratory (Figure 1.8). They recreated the primitive atmosphere consisting of water, hydrogen, methane and ammonia. Sparks were discharged in the synthetic atmosphere to mimic lightning, a condenser cooled the atmosphere, raining water and dissolved compounds back to the miniature sea. As materials circulated through the apparatus, the solution in the flask changed from clear to murky brown. After one week, when the contents of the solution were analyzed a variety of organic compounds, including urea, organic acids, sugars, nitrogenous bases and amino acids (glycine, alanine and aspartate) were synthesized (Figure 1.9).

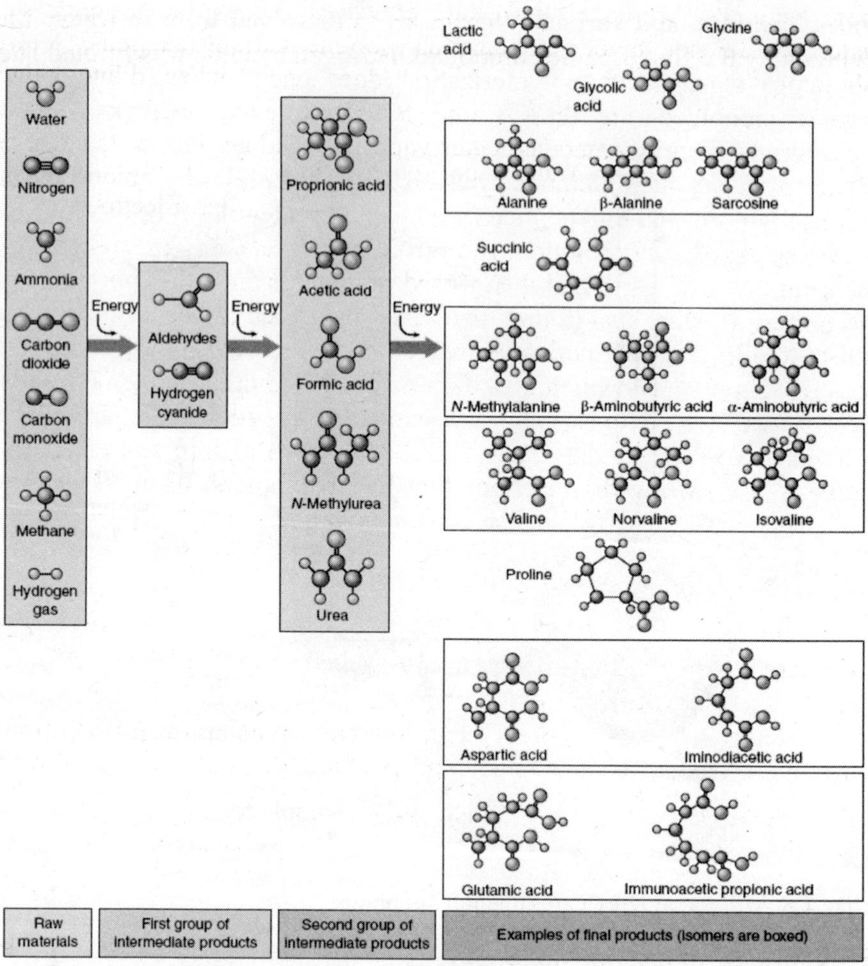

Figure 1.9 . Results of the Miller-Urey experiment. Seven simple molecules, all gases, were included in the original mixture. Note that oxygen was not among them; instead, the atmosphere was rich in hydrogen. At each stage of the experiment, more complex molecules were formed: first aldehydes, then simple acids, then more complex acids. Among the final products, the molecules that are structural isomers of one another are grouped together in boxes. In most cases only one isomer of a compound is found in living systems today, although many may have been produced in the Miller-Urey experiment.

1.3.5 Synthesis of organic polymers

After abiotic synthesis of organic monomers, the next step was aggregation of these monomers to form polymers (Polypeptides, polysaccharides, nucleic acids). Today synthesis of these polymers takes place inside the cell and it is considered as an anabolic process which requires energy in the form of ATP and such reactions are catalyzed by enzymes, but in primitive environmental conditions abiotic synthesis of these polymers would have had to occur without the help of these enzymes. Moreover the dilute concentration of these monomers in sea water would not favor spontaneous polymerization reactions. However in laboratory experiments polymerization does occur when dilute solutions of organic monomers are dipped into hot sand, clay, or rocks. This process concentrates the monomers as water vaporizes. Sidney Fox of the University of Miami by using such methods synthesized polypeptides by abiotic means. Such abiotically produced polypeptides were called protenoids by Fox. Perhaps waves or rain splashed dilute solutions of organic monomers onto fresh lava or other hot rocks on the early earth and then rinsed protenoids and other polymers back into the water. It was also found that clay has been especially important as a substratum for the polymerization reactions – a prerequisite to life. Clay particles possess charged sites which assist amino acids and other monomers to bind with electrostatic interactions and thus concentrates monomers from the dilute solutions. At some binding sites, metal ions, such as iron and zinc, function as catalysts, facilitating the reactions that link monomers. Even today these metal ions play an important role in catalysis in metalloenzymes.

1.3.6 Aggregation of organic polymers

Abiotically synthesized polymers aggregate to form *protobionts.* These are considered to be the first life form and living cells may have been preceded by protobionts. They exhibit some properties of life, like they are capable of precise reproduction and can maintain their internal environment different from their surrounding as well as metabolism and excitability.

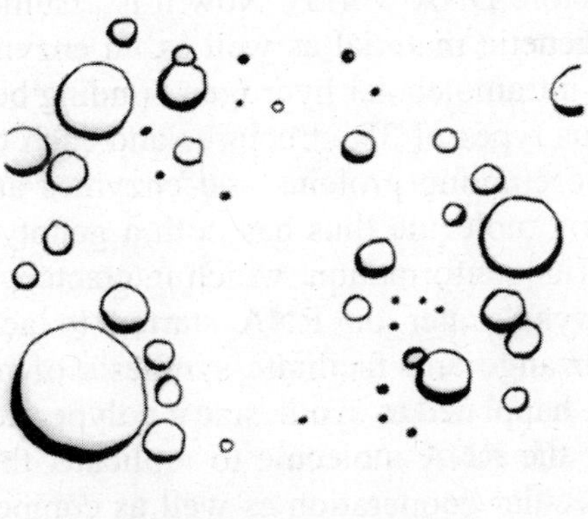

Figure 1.10a. Coacervates

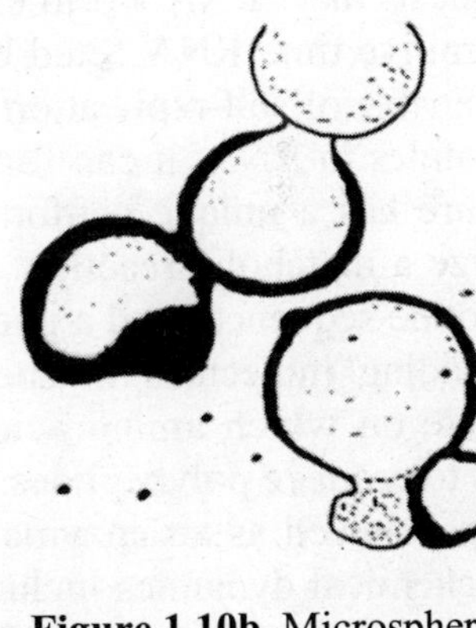

Figure 1.10b. Microspheres

One type of protobiont is **coacervate** named by Oparin. It is a stable droplet that tends to self-assemble when a suspension of macromolecules (polypeptides, nucleic acids, and polysaccharides) is shaken (Figure 1.10a). Coacervates are surrounded and stabilized by a shell of water molecules. If enzymes are incorporated in the coacervates then they are able to absorb substrates from their surroundings and release the products of the reaction catalyzed by the enzymes. Another type of protobiont is **microsphere** (Figure 1.10b) named by Sidney Fox.

They are formed when water is added to thermal protenoids. They are coated by a protein membrane which is selectively permeable. Microspheres undergo osmotic swelling or shrinkage when placed in solutions of different salt concentrations. Some microspheres also store energy in the form of **membrane potential.** They can also discharge potential in a nerve-like fashion. Such excitability is characteristic of all life forms.

1.4 RNA MAY HAVE BEEN THE FIRST GENETIC MATERIAL

After synthesis of the first life form i.e. **protobionts** the major thing was perpetuation of these protobionts. These droplets uptook nutrients in the form of abiotically synthesized monomers from the surrounding environment (sea water) and divide to give produce their progeny. As these droplets grew split, grew, and split again, their unique catalysts and other functional biomolecules were increasingly diluted. Natural selection favors droplets which develops the mechanism to copy their biomolecules, some mechanism of heredity, and then distribute equally before splitting to give the progeny cells. Since we know that metabolism is a key feature of life and all metabolic processes are governed by enzymes present inside the cell, the information regarding synthesis of enzymes lies in the genetic material DNA in the modern cells. It was experimentally proven that even protobionts possess metabolic activity; it means that protobionts have some biomolecules having catalytic activity. If these biomolecules were enzymes than there must have been DNA present to encode them. After the discovery of RNA acting as an enzyme (ribozyme), it gives strong support to the hypothesis that "RNA world existed before DNA world". Now it is assumed that in primitive time RNA acted both as genetic material as well as an enzyme and was capable of self-replication. Due to intramolecular hydrogen bonding between nucleotides in RNA, it can form various types of 3D structures and each type of structure has a unique conformation resembling proteins and enzymes and can catalyze a metabolic reaction. The RNA molecule thus has both a genotype (its nucleotide sequence) and a phenotype (its conformation, which interacts with the surrounding molecules in specific ways). Later on RNA started to act as a template on which amino acids can arrange and facilitate synthesis of peptide bonds to produce polypeptides. If RNA happened to synthesize a polypeptide that in turn behaved as an enzyme helping the RNA molecule to replicate, then the early chemical dynamics included molecular cooperation as well as competition. This first step towards the replication and translation of genetic information may have been taken by molecular evolution even before RNA and polypeptides became packaged within membranes (Figure 1.11a). The packaging of primitive RNA and their polypeptide products within a membrane would have been a significant milestone in the early history of life and polypeptides facilitate replication of RNA enclosed in limiting membrane rather than competing RNA molecules in its environment, i.e. external to limiting membrane (Figure 1.11b).

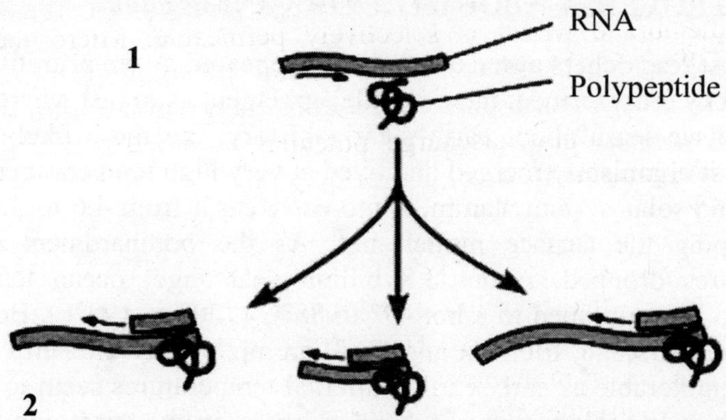

(a) A molecular "free-for-all" in the prebiotic soup of organic molecules.

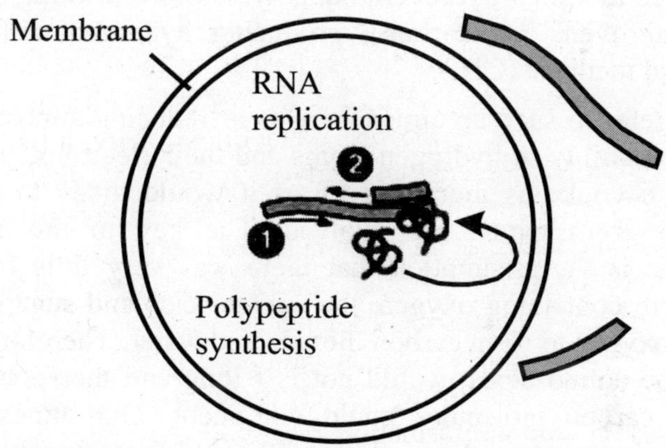

(b) Exclusive cooperation among membrane-enclosed molecules.

Figure 1.11. Hypotheses for the beginning of molecular cooperation

It is well understood that RNA is chemically less stable than DNA because it is rapidly hydrolyzed even if there is slight pH fluctuation, therefore, RNA cannot be considered as a stable genetic material. Meanwhile in some primitive cells ribose sugar gives rise to deoxyribose sugar and uracil gives rise to thymine and this change inside the cell is responsible for synthesis of DNA. As DNA is chemically more stable than RNA, DNA started to act as the genetic material and RNA started to act as intermediatory in transferring the message given by DNA to proteins and enzymes and such types of cells are more stable as compared to cells having RNA as genetic material and out competes them by utilizing more resources from the surrounding environment. Such cells are selected by natural selection and slowly only those cells survive which are having DNA as a genetic material.

1.5 CONTROVERSY ABOUT THE ORIGIN OF LIFE

While most researchers agree that life first appeared as the primitive earth cooled and its rocky crust formed, there is little agreement as to just where this occurred. The more we learn about earth's early history, the more likely it seems that earth's first organisms emerged and lived at very high temperatures. Rubble from the forming solar system slammed into early earth from 4.6 to 3.8 billion years ago, keeping the surface molten hot. As the bombardment slowed down, temperatures dropped. About 3.8 billion years ago, ocean temperatures are thought to have dropped to a hot 49° to 88°C (120° to 190°F). Between 3.8 and 3.5 billion years ago, life first appeared, promptly after the earth was habitable. Thus, as intolerable as early earth's infernal temperatures seem to us today, they gave birth to life. Very few geochemists agree on the exact composition of the early atmosphere. One popular view is that it contained principally carbon dioxide (CO_2) and nitrogen (N_2) along with significant amounts of water vapor (H_2O). It is possible that the early atmosphere also contained hydrogen (H_2) and compounds in which hydrogen atoms were bonded to the other lighter elements (sulfur, nitrogen, and carbon), producing hydrogen sulfide (H_2S), ammonia (NH_3), and methane (CH_4).

We refer to such an atmosphere as a 'reducing atmosphere' because of the ample availability of hydrogen atoms and their electrons. In such an atmosphere it would not take as much energy as it would today to form the carbon-rich molecules from which life evolved. The key to this reducing atmosphere hypothesis is the assumption that there was very little oxygen around. In an atmosphere containing oxygen, the amino acids and sugars react spontaneously with the oxygen to form carbon dioxide and water. Therefore, the building blocks of life, the amino acids, would not last long and the spontaneous formation of complex carbon molecules could not occur. Our atmosphere changed once organisms began to carry out photosynthesis, harnessing the energy in sunlight to split water molecules and form complex carbon molecules, giving off gaseous oxygen molecules in the process. The earth's atmosphere is now approximately 21% oxygen.

Critics of the reducing atmosphere hypothesis point out that no carbonates have been found in rocks dating back to early earth. This suggests that at that time carbon dioxide was locked up in the atmosphere, and if that was the case, then the prebiotic atmosphere would not have been reducing. Another problem of the reducing atmosphere hypothesis is that because a prebiotic reducing atmosphere would have been oxygen-free, there would have been no ozone. Without the protective ozone layer, any organic compounds which might have formed would have been broken down quickly by ultraviolet radiation.

Laboratory simulation cannot prove that the kind of chemical evolution that has been described in this chapter created life on the primitive earth, but only some of the key steps could have happened. The origin of life remains a matter of scientific speculation, and there are alternative views of how several key processes occurred. It is argued whether the abiotic synthesis of organic monomers occurred on primitive earth. It is possible that some organic compounds reached the early earth from space. This idea called **panspermia**, explains that hundreds of thousand of meteorites and comets hitting the early earth brought with them organic molecules formed abiotically in outer space. Extraterrestrial organic compounds, including amino acids, have been found in modern meteorites, and it seems that these bodies could have enriched the early earth with organic compounds. Both panspermia and chemical evolution could have contributed to the pool of organic molecules that formed the earliest life, but many scientists believe that extraterrestrial sources made only a minor contribution.

OTHER SUGGESTIONS

If life did not originate at the ocean's edge under the blanket of a reducing atmosphere, where did it originate? Various hypotheses are discussed below:

Under frozen oceans: One hypothesis proposes that life originated under a frozen ocean, not unlike the one that covers Jupiter's moon Europa today. All evidence suggests, however, that the early earth was quite warm and this hypothesis seems unlikely.

Deep in the earth's crust: Another hypothesis is that life originated deep in the earth's crust. In 1988 Gunter Wachtershauser proposed that life might have formed as a by-product of volcanic activity, with iron and nickel sulfide minerals acting as chemical catalysts to recombine gases spewing from eruptions into the building blocks of life. In later works he and his coworkers were able to use this unusual chemistry to build precursors for amino acids (although they did not actually succeed), and to link amino acids together to form peptides. Critics of this hypothesis point out that the concentration of chemicals used in their experiments greatly exceed what is found in nature.

Within clay: Other researchers have proposed the unusual hypothesis that life is the result of silicate surface chemistry. The surface of clay have positive charges to attract organic molecules, and exclude water, providing a potential catalytic surface on which life's early chemistry might have occurred. While interesting conceptually, there is little evidence that this sort of process could actually occur.

At deep-sea vents: Becoming more popular is the hypothesis that life originated at deep-sea hydrothermal vents, with the necessary prebiotic molecules being synthesized on metal sulfides in the vents. The positive charge of the sulfides would have acted as a magnet for negatively charged organic molecules. In part, the current popularity of this hypothesis comes from the new science of genomics, which suggests that the ancestors of today's prokaryotes are most closely related to the bacteria that live on the deep-sea vents.

The universe has 10^{20} (100,000,000,000,000,000,000) stars similar to our sun. We don't know how many of these stars have planets, but it seems increasingly likely that many do. Since 1996, astronomers have been detecting planets orbiting distant stars. At least 10% of the stars are thought to have planetary systems. If only 1 in 10,000 of these planets is the right size and at the right distance from its star to duplicate the conditions in which life originated on earth, the "life experiment" will have been repeated 10^{15} times (that is, a million billion times). It does not seem likely that we are alone.

Ancient Bacteria on Mars?

A dull gray chunk of rock collected in 1984 in Antarctica ignited an uproar about ancient life on Mars with the report that the rock contains evidence of possible life. Analysis of gases trapped within small pockets of the rock indicated it is a meteorite from Mars. It is, in fact, the oldest rock known to science–fully 4.5 billion years old. Back then, when this rock formed on Mars, that cold and arid planet was much warmer, flowed with water, and had a carbon dioxide atmosphere–conditions not too different from those that spawned life on earth.

When examined with powerful electron microscopes, carbonate patches within the meteorite exhibit what look like microfossils, some 20 to 100 nanometers in length. One hundred times smaller than any known bacteria, it is not clear they actually are fossils, but the resemblance to bacteria is striking.

Viewed as a whole, the evidence of bacterial life associated with the Mars meteorite is not compelling. Clearly, more painstaking research remains to be done before the discovery can claim a scientific consensus. However, while there is no conclusive evidence of bacterial life associated with this meteorite, it seems very possible that life has evolved on other worlds in addition to our own.

Other Planets

Has life evolved on other worlds within our solar system? There are planets other than ancient Mars with conditions not unlike those on earth. Europa, a large moon of Jupiter, is a promising candidate (Figure 1.12). Europa is covered with ice, and photos taken in close orbit in the winter of 1998 reveal seas of liquid water beneath a thin skin of ice. Additional satellite photos taken in 1999 suggest that a few miles under the ice lies a liquid ocean of water larger than earth's, warmed by the push and pull of the gravitational attraction of Jupiter's many large satellite moons. The conditions on Europa now are far less hostile to life than the conditions that existed in the oceans of the primitive earth. In coming decades satellite missions are scheduled to explore this ocean for life.

Where life began is another issue. Till now it is considered that the shallow waters and moist sediments are the most likely sites for the origin of first life. Molecular phylogenetic analyses indicate that the ancestors of modern prokaryotes thrived in very hot conditions and may have lived on inorganic sulfur compounds that are common in deep sea vents.

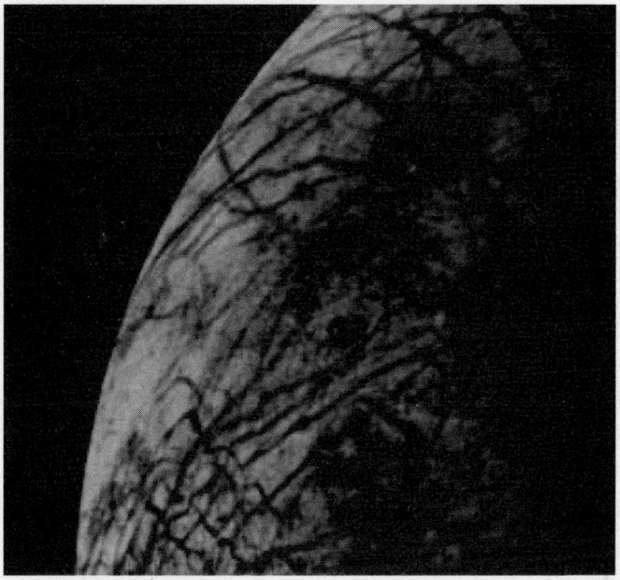

Figure 1.12. Is there life elsewhere? Currently the most likely candidate for life elsewhere within the solar system is Europa, one of the many moons of the large planet Jupiter.

■■■■

Chapter 2

EVOLUTIONARY THEORIES AND EVIDENCES IN FAVOR OF EVOLUTION

2.1 Lamarckism
    2.1.1 Main postulates of Lamarckism
    2.1.2 Criticism
2.2 Darwinism
    2.2.1 Main postulates of Darwinism
    2.2.2 Artificial selection
    2.2.3 Sexual selection
    2.2.4 Critics of Darwinism
2.3 The Modern synthesis
2.4 Evidences in favor of evolution
    2.4.1 On small scale: Natural selection provides evidence of evolution
        2.4.1.1 The evolution of insecticide-resistant insects
        2.4.1.2 The evolution of drug-resistant HIV
    2.4.2 Evidences of large scale evolutionary changes
        2.4.2.1. Evidences from Biogeography
        2.4.2.2. Evidences from Fossils
        2.4.2.3. Geological time scale
        2.4.2.4. Evidences from comparative anatomy
        2.4.2.5. Evidences from embryology
        2.4.2.6. Evidences from Biochemistry and molecular biology

The history of evolutionary biology is very old. A number of Greek philosophers had ideas about the gradual evolution of life. But the philosophers who influenced the western culture most, Plato (427-347 BC) and his student Aristotle (384-322 BC), held opinions that opposed any concept of evolution. They believed that species are permanently fixed in form (remain unchanged), are perfect and do not evolve. In the 18th Century with the concept of theories of evolution of Lamarck and Darwin, the concept of fixity of species was changed. Both developed a comprehensive model to explain how one species evolves into another (transformation of species).

2.1 LAMARCKISM

Figure 2.1. Jean-Baptiste Lamarck

Jean-Baptiste Lamarck (1744- 1829), (Figure 2.1), a French naturalist, was in charge of the invertebrate collection at the Natural History Museum in Paris. He published his work *Philosophie Zoologique* (1809), the year Darwin was born, in which he argued that species change over time into new species. The way in which he thought species changed was importantly different from Darwin's and our modern idea of evolution. The word "transformism" was commonly used to describe Lamarck's idea. Figure 2.2 describes Lamarck's conception of evolution, and how it differs from Darwin's and our modern concept. Lamarck believed that lineages of species persisted indefinitely, changing from one form into another; lineages in his system did not branch and did not go extinct.

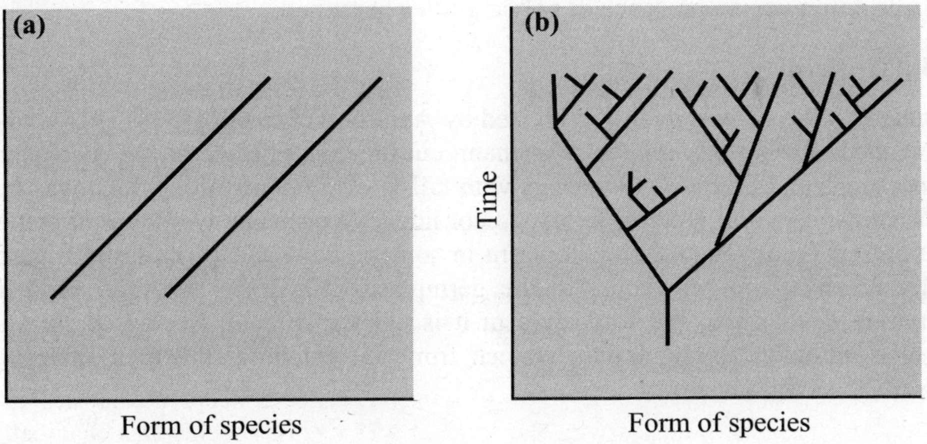

Figure 2.2. (a) Lamarckian "transformism". (b) Darwinian evolution is tree-like, as lineage split, and allows for extinction.

2.1.1 Main Postulates of Lamarckism

1. Internal vital forces

According to Lamarck, every organism has some internal vital forces by which he can develop new structures/organs according to its own will.

2. Environment fluctuation and new needs

Environment is dynamic in nature. Whenever it changes, it desires new needs to fulfill new needs an organism develop new structures/organs by using its internal vital forces.

3. Use and Disuse of organs

According to Lamarck if an organ is being constantly used it becomes more developed and disuse of an organ leads to degeneration of that organ

4. Inheritance of acquired characters

During the life time of an organism due to internal vital forces, environmental changes and use and disuse of an organ the organism acquires new organs/characters. Such acquired characters are inherited to the next generation and after inheritance of such characters through successive generation, give rise to generation of new species.

Lamarck put forward some examples to justify his theory: Giraffes developed a long neck and long forelimbs to eat leaves from tall trees. He also cited that a blacksmith develops bigger biceps because he wields a hammer constantly. There is, however, no evidence that acquired characteristics can be inherited. The blacksmith may increase strength and stamina by a life time of pounding with a heavy hammer, but these acquired traits do not change genes transmitted by gametes to offspring. He also gives some examples of vestigial organs, for example, presence of pelvic girdles in python.

2.1.2 Criticism

Lamarck's theory was mainly criticized by August Weismann (1834–1914), who gave the *theory of germplasm.* Weismann cut the tails of mice for 22 successive generations and he found offsprings with tail in every generation which was not in accordance with Lamarck theory. According to Weismann, two types of matter are present in organisms, somatoplasm in somatic cells and germplasm in germ cells. If any variation occurs in the germplasm it will be inherited while if variation develops in the somatoplasm it is not transmitted. Boring of ear and nose is common among Indian women from ancient times but their offspring never shows this character.

2.2 DARWINISM

After the doubt about the permanence of species were beginning to form and it got a striking support from the Darwinian Revolution. Charles Darwin (1809-1882) was born in Shrewsbury in western England. Right from his childhood he took keen interest in nature (Figure 2.3).

Figure 2.3. Charles Darwin in 1849 at the age of 40

Darwin's father, an eminent physician, sent Charles Darwin to the University of Edinburg to study Medicine. But Darwin found medical school boring and distasteful. He left Edinburg without a degree and shortly thereafter enrolled at Christs College at Cambridge University, where he received his B.A. degree in 1831. The crucial period of his life, from when he was 22 years old, had traveled the world as a naturalist aboard *HMS Beagle* (1832-1837). On the voyage of Beagle (Figure 2.4) he spent most of his time on shore, observing and collecting thousands of specimens of South American plants and animals. He observed the various adaptations of plants and animals of the Brazilian jungles, grasslands of the Argentine pampas, the desolate lands of Tierra del Fuego near Antartica, and the Andes mountains. He noticed that plants and animals of temperate regions of South America was more closely related to species of tropical region of South America than to species in temperate regions of Europe. Among the birds Darwin collected were several types of finches from different islands of Galapagos variations in their beaks, based on their food habit. He had initially supposed that the Galapagos finches all belonged to one species, but it now became clear that each island had its own distinct species (Fig. 2.5). His observations of geographic variation probably led Darwin to accept that species can change. The next important step was to give a scientific theory to explain why species change. Darwin expounded several ideas, including Lamarckism, but rejected them all because they failed to explain a crucial fact – adaptation. Soon after returning to Great Britain in 1836, Darwin started reassessing all that he had observed during his voyages.

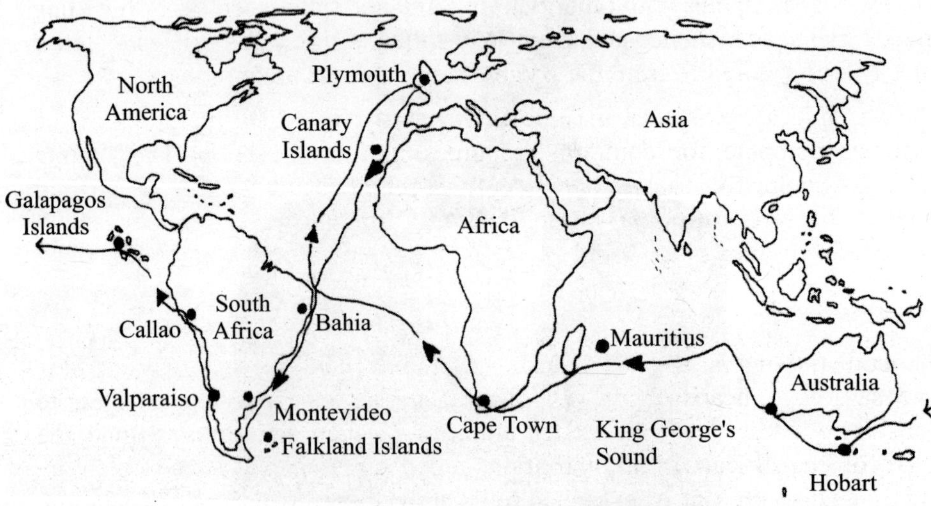

Figure 2.4. Route of the 5-year voyage on the S.S. Beagle.

By the early 1840s, Darwin had worked out the major features of his theory of natural selection as the mechanism of evolution. However, he had not yet published his ideas. In 1844, Darwin wrote a long essay on the origin of species and natural selection. But he did not present it publically due to possible controversy. In June 1858, Darwin received a letter from Alfred Russel Wallace (1823-1913), a young British naturalist working in the East Indies. The letter was accompanied by a manuscript in which Wallace developed a theory of natural selection essentially identical to Darwin's. Darwin's friends Charles Lyell and Joseph Hooker, arranged for a simultaneous announcement of Darwin and Wallace's ideas at the Linnaean Society in London in 1858. Darwin quickly finished *The Origin of Species* and published it the next year. Although Wallace wrote up his ideas for publication first, Darwin developed and supported the theory of natural selection so much more extensively that he is known as its main architect, and Darwin notebook prove that he formulated the theory of natural selection 15 years before reading Wallace's manuscript. *The Origin of Species* mainly emphasizes two main points: the occurrence of evolution and natural selection. The main features of Darwin's theory can be explained as follows:

2.2.1 Main postulates of Darwinism

1. Over-production

All organisms have a natural tendency to produce more and more offsprings i.e. organisms multiply in geometric ratio, for example, plants produce thousands of seeds, an insect lays hundreds of eggs and if all the offspring's survive and grow than it will lead to over production.

2. Struggle for existence

Every individual competes with others of the same and other species for fulfilling basic needs like space, shelter and food. It is called struggle for existence and it continues for the whole life from the zygote stage upto its natural death.

The struggle for existence may be **intraspecific** in which members of the same species compete for common resources like food, shelter and mating partner. In **Interspecific** struggle members of different species competes for food and shelter. In **environmental struggle** every organism has to cope with environmental extremes like cold, heat, wind, rain, drought, flood and epidemic.

3. Variations

Except identical twins no two individuals look similar and their requirements are also not same. It means there are differences among individuals. Due to variations some individuals are better adjusted to their surrounding than the others. According to Darwin, the variations are continuous and those which are helpful in the adaptation of an organism to its surrounding would be passed on to the next generation, while the others disappear.

4. Natural selection

In the struggle for existence individuals which are more fit are successful. The individuals with more favourable adaptations will live a longer life and are able to win over their mating partner. Darwin called it ***sexual selection***. In the struggle for existence only those members survive which possess useful variations means nature selects fit individuals. This was called ***natural selection*** and it operates by differential survival and differential reproduction.

5. Origin of new species

Darwin explained that variations appearing due to environmental changes are transmitted to the next generation. So offsprings become different from their ancestors. In the next generation, process of natural selection repeats so after many generations a new species is formed.

In addition to natural selection Darwin also wrote about artificial selection, sexual selection and he presented the theory of pangenesis to explain inheritance.

2.2.2 Artificial selection

Darwin explained the power of selection as a force in evolution with examples from artificial selection. Man has been taking the advantage of genetic variations for improving the quality of domesticated plants and animals. He selects the individuals with desired characters and separates them from those which do not have such characters. This process is termed as artificial selection. Thus this process is manmade. If it is repeated for many generations it produces a new breed with the desired characters. By artificial selection animal breeders are able to produce improved varieties of domestic animals like dogs, horse, pigeons, poultry cows, goats, sheep and pigs from their wild ancestors. Similarly, the plant breeders have obtained improved varieties of useful plants like wheat, rice, sugarcane, cotton, pulses, vegetables, fruits etc. Artificial selection is similar to natural selection except that the role of nature is taken over by man and the characters selected are of human use.

2.2.3 Sexual selection

The male and females of an animal species obviously differ in their reproductive organs. But in many species, there are also marked differences, called secondary sexual characteristics, which are not directly associated with reproduction. This distinction in appearance is called as sexual dimorphism. It often manifested in size difference, usually one in which males are larger than the females. But sexual dimorphism can also be evident in such features as colorful plumage of male birds, manes on male lions, antlers on male deer etc.

Sexual dimorphism is a product of what Darwin called sexual selection. Certain evolutionary traits like horns, antlers, plumage in males can be explained by sexual selection. Today biologists call sexual selection as intrasexual selection and intersexual selection. **Intrasexual selection** is the direct competition among individuals of one sex (usually males in vertebrates) for mates of the opposite sex. Males may use secondary sexual equipment such as antlers to battle with competitors. This is especially common in species where a single male garners a harem of females. These males may gain their status by defeating smaller, weaker, or less fierce males in combat. In **intersexual selection**, also called as mate choice, usually females choose mates of their own choice and the traits selected by such mate choice are called ornaments (plumage in male birds). Genes that enable male to develop impressive ornaments or fighting abilities helps male to mate with female and thus transmit such genes in next generation (sexual selection).

According to Charles Darwin beards in human males and the relative hairlessness of human compared to nearly all mammals are a result of sexual selection.

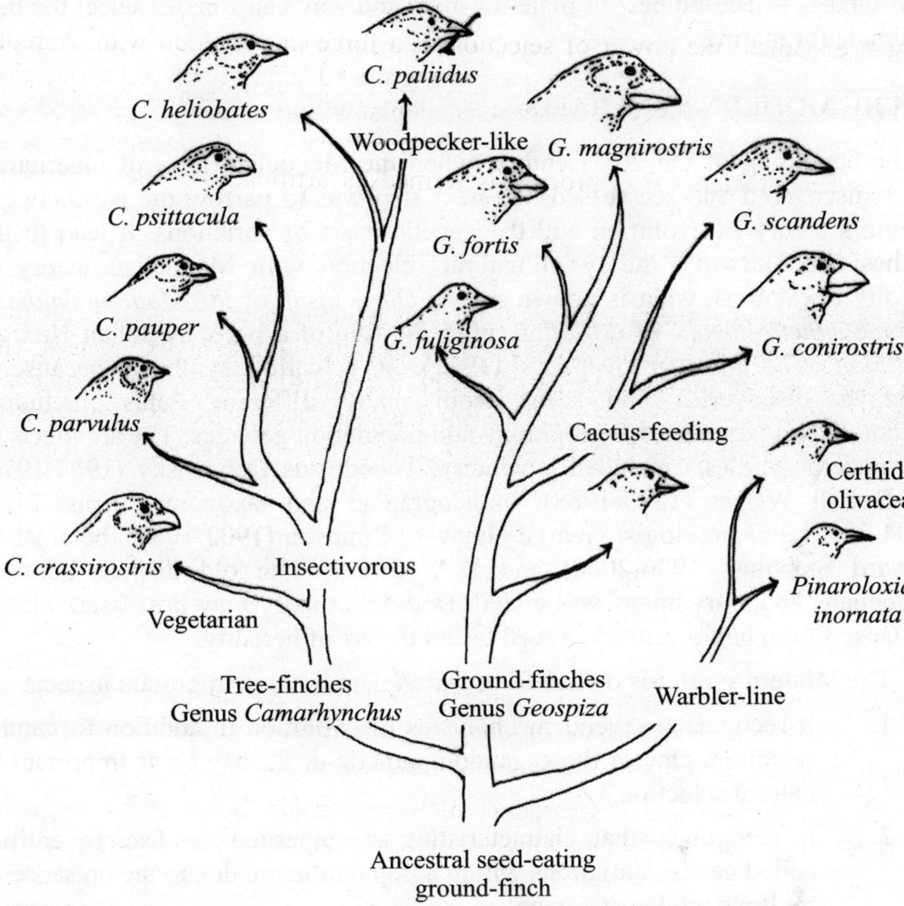

Figure 2.5. Evolutionary tree of Darwin's finches showing beak adaptations of the individual species.

2.2.4 Critics of Darwinism

There was a mixed response to Darwin's theory. The idea of evolution become controversial and it seemed to contradict the Bible, in which the various kinds of living things are said to have been created separately. But professional biologists defended the new evolutionary view against religious attack. One of the major objection to Darwin's theory was that he was unable to explain the cause and origin of variations because it lacks satisfactory theory of heredity. He even produced a theory of heredity called as "Pangenesis". Pangenesis holds that body cells shed gemmules which collect in the reproductive organs prior to fertilization. Thus every cell of the body contributes in the constitution of the offsprings. Pangenesis itself is now seen as deeply flawed and not supported by any observation.

Darwin was also criticized on his view regarding sexual selection. His view is applicable only to small numbers of population. It does not fit in mind why only females select the best fit male for mate and why can't males select the best fit female for mating.

2.3 THE MODERN SYNTHESIS

At the beginning of the 20th Century, when the Mendelian laws of inheritance was rediscovered and reassessed, the next step was to narrow the gap between Darwin's theory of evolution and the genetic basis of variations. It lead to the synthesis of Darwin's theory of natural selection with Mendelian theory of heredity to establish what is known as *neo-Darwinism*, or *the Modern synthesis* or the *Synthetic theory of evolution,* after the title of a book by Julian Huxley, *Evolution: The Modern Synthesis* (1942). It is called synthesis because it integrates discoveries and ideas from many different fields, including paleontology, taxonomy, biogeography and population genetics. The architects of this modern synthesis included geneticists Theodosius Dobzansky (1900-1975) and Sewall Wright (1889-1988), biogeographer and taxonomist Ernst Mayr (1904-2005), paleontologist George Gayward Simpson (1902-1984), botanist G. Ledyard Stebbins (1906-2000) and R.A. Fisher. The old dispute between Mendelians and Darwinians was ended. Darwin's theory now possessed what it had lacked for a half a century; a well tested theory of heredity.

The Modern synthesis differs from Darwinism in three important aspects:

1. It recognizes several mechanisms of evolution in addition to natural selection. One of these, random genetic drift, may be as important as natural selection.
2. It recognizes that characteristics are inherited as discrete entities called genes. Variations within a population are due to the presence of multiple alleles of a gene.
3. It postulates that speciation is (usually) due to the gradual accumulation of small genetic changes over long periods of time which leads to large changes, *i.e.* macroevolution is simply a lot of microevolution.

The modern synthesis helps us to understand the mechanism of evolution and speciation in a simple way as follows:

Overproduction

All organisms have a natural tendency to multiply in geometric ratio.

Struggle for existence

It is the struggle for common resources like food, shelter and mating partner. This struggle may be intraspecific or interspecific. All individuals of various populations have to cope with environmental extremes in the environmental struggle.

Variations

These are inheritable variations which can occur due to the following reasons:

a. **Mutations:** These are discontinuous variations which develop due to change in physical and chemical properties of the genome.

   Gene mutation: These are point mutation which changes nucleotide sequence of the gene either by transition or transversion type of event.

   Chromosomal mutation: These are numerical change (polyploidy/aneuploidy) or structural change in chromosomes.

b. **Genetic drift**: It is chance fluctuation in the gene pool of a small population

c. **Genetic recombination**: It is continuous source of variation due to crossing over in sexually reproducing organisms.

Natural selection

It is differential survival and differential reproduction in population. Individual having variations which are more adaptive in value will live a longer life and reproduce many times in their life time and thus contribute more offsprings in their next generation. In this way natural selection always operates in every generation and selects the best genotype.

Reproductive isolation

Reproductive isolation is a segregation of population by some barriers which prevent interbreeding and leads to formation of new species.

Origin of new species

An isolated population of a species independently develops different types of variations (due to genetic changes). They later accumulate in its gene pool. After several generations the isolated population becomes genetically and reproductively different from others so as to constitute a new species.

2.4 THE EVIDENCES FOR EVOLUTION

Evidences for evolution are strong proof in support for the theory of evolution, that it is a continuous process which had occurred in the past from the time of the origin of life and is still continuing and will shape the diversity of life in future. On small scale there are some evidences of rapid evolutionary changes which can be directly observed. However we can also observe large-scale evolutionary changes of biological diversity of hundreds of millions of years. We will first look at small-scale evolutionary changes by taking examples of natural selection as an evidence of evolution.

2.4.1 On Small Scale: Natural Selection provides evidence of Evolution

We will take two examples of natural selection as a pervasive mechanism of evolution in populations:

The evolution of insecticide-resistant insects

A classic example of natural selection is the evolution of insecticide resistance in hundreds of insect species. Insecticides are poisons used to kill insects that are pests in farmlands, swamps and homes. Examples are DDT, now banned in many countries, and malathione. The poison kills most of the members of the insect population, leaving the resistant individuals to reproduce and their offspring inherit the genes for insecticide resistance. In each generation, the proportion of insecticide-resistant individuals in the insect population increases. The population has adapted to change in its environment.

This example of insect adaptation to insecticides highlights two key points about natural selection. First, notice that natural selection is more a process of editing than it is a creative mechanism. An insecticide does not create resistant individuals, but selects for resistant insects that were already present in the population. Second, natural selection works according to time and place. It favors those characteristics in a varying population that fit the current, local environment. Characteristics which are adaptive in one situation may be useless or even detrimental in other circumstances.

The evolution of drug-resistant HIV

Human immunodeficiency virus (HIV) that causes AIDS is a retrovirus having RNA as its genetic material. Inside the host cell by using its own reverse transcriptase enzyme it synthesizes DNA version from its RNA version. The DNA synthesized in the process of reverse transcription is then inserted into the host cell genome and this condition is called as **prophage**. In later stages it multiplies by using transcriptional machinery of the host cell to synthesize its RNA component. Reverse transcriptase is normally not present in human cells. The reverse transcriptase is the favorite target for anti-HIV drugs. If reverse transcriptase can be inactivated by a drug, the virus is stopped from reproducing without any damaging side effects on the cell.

Many drugs have been developed against reverse transcriptase. One large class of these drugs consists of nucleoside inhibitors. A nucleoside is a nucleotide without the phosphate, it is base plus sugar. The drug 3TC, for example, is a molecule which mimics the C (cytosine) nucleotide of DNA. The HIV's reverse transcriptase will pick up a 3TC molecule instead of C and insert it into a growing DNA chain. This error terminates further elongation of the DNA and thus blocks reproduction of the HIV.

When AIDS patients are treated with 3TC, initially the HIV population in the human body decreases by a huge amount. But then, within days, 3TC-resistant strains of HIV start to be detected. The drug-resistant HIV then increases in frequency, and within 3 weeks of the start of the drug treatment the population of drug-resistant HIV increases up to 100% in the patients body. The change, from a viral population that was susceptible to 3TC to a viral population that was resistant to 3TC, is an example of evolution by natural selection. The evolution takes place within a single human body, and is exceptionally rapid relative to most examples of evolution.

The evolution of drug resistance can be understood at molecular level. The change from 3TC-susceptible HIV to 3TC-resistant HIV is achieved by a change in one codon in the gene that codes for reverse transcriptase. The amino acid methionine is changed to one of the three other amino acids. The methionine is in part of the reverse transcriptase that interacts with the nucleotides. Due to this change the 3TC-resistant variety of HIV has a slightly different version of reverse transcriptase that is able to discriminate between the drug and the normal C nucleotide. Members of the resistant HIV population that inherit the gene for this form of the enzyme have no advantage in the absence of 3TC, in fact, they replicate their DNA more slowly than the normal variety of HIV. But once 3TC is added to the environment of these viruses, it becomes a potent force in natural selection, favoring reproduction of the resistant individuals.

2.4.2 Evidences of large-scale evolutionary changes

We have examined cases of evolution by natural selection that occur rapidly enough to be directly observed. However, the much grander changes of biological diversity occur on a time scale spanning hundreds of millions of years. For such large-scale evolutionary changes human observation is too short-term to witness the whole history of life, but their are some reliable evidences from biogeography, paleontology, comparative embryology, homologies, analogies, biochemistry, molecular biology and taxonomy which has witnessed long-term evolutionary changes. Among these evidences biogeographical and paleontological evidences are considered as the best evidences in favor of evolution.

Evidence from biogeography

The study of geographical distribution of animals and plants in different part of the earth is called biogeography. It was biogeography, even more than fossils that first nudged Darwin and Wallace towards the evolutionary view of life. Darwin noticed that species in one biogeographic area tend to be relatively similar. Species living in different areas tend to differ more, even if the species occupy a similar ecological niche. Thus ecologically different species in one area will share similarities that are lacking between ecologically similar species in different areas. This suggests the species in any one area are descended from a common ancestor. For example, Australia is the home of subgroups of mammals- the Prototherians and marsupials which are distinct from another group of mammals – the eutherians that lives elsewhere on earth. Eutherians are mammals that complete their embryonic development in the uterus, while marsupials are born as embryos and complete their development in an external pouch, and prototherians are egg, laying mammals like Platypus and Echidina. For example, a forest dwelling marsupial called the sugar glider is superficially very similar to flying squirrels, eutherians that live in North American forests. These two mammals have adapted to the same way of life, but they evolved independently from different ancestors. The sugar glider is distinctly marsupial, much more closely related to kangaroos and other Australian marsupials than to flying squirrels or any other eutherian mammals. The sugar glider is a marsupial not because that is a requirement for its gliding lifestyle but simply because its ancestors were marsupials. The unique fauna of Australia diversified on that island continent after it became isolated from the land masses on which placental mammals diversified. The resemblance between sugar gliders and flying squirrels is an example of what biologists call convergent evolution.

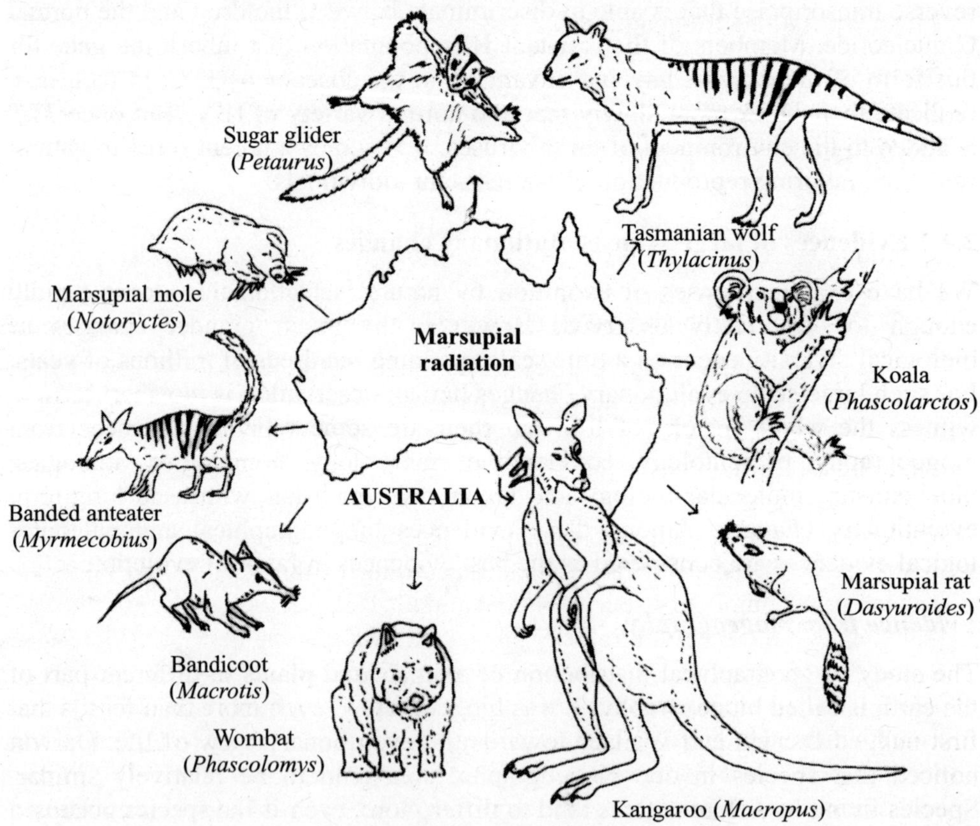

Figure 2.6. Adaptive radiation of Australian marsupial mammals showing the many divergent forms that evolved independently of changes occurring among placental mammals on other continents.

Islands are the showcases of biogeographic evidences for evolution. They generally have many species of plants and animals that are endemic. And yet, as Darwin observed when he reassessed his collections from the voyages on the *Beagle*, most island species are closely related to species from the nearest mainland or neighboring islands. This explains why two islands with similar environments in different parts of the world are populated not by closely related species but by species taxonomically affiliated with the plants and animals of the nearest mainland, where the environment is quite different. Island chains or archipelagos are especially interesting in their biogeography. If a species that disperses from a mainland to an island succeeds in its new environment, it may give rise to several new species as populations spread to other islands in the archipelago. The name given to this process, **adaptive radiation**, signifies the rapid evolution of one or a few forms into many different species that occupy different habitats within a new geographical area. The evolution of Darwin finches on Galapagos Islands (Figure 2.5) and the marsupial radiation in Australia (Figure 2.6) are good examples of adaptive radiations.

Theory of Island Biogeography

Besides reflecting the past and present distribution of species biogeography also provides valuable information about species diversity of a geographical region particularly islands.

Because of their limited size and isolation, islands provide excellent opportunities for studying some of the factors that affect the species diversity of communities. By "islands" we mean not only oceanic islands but also habitat islands on land, such as lakes, mountain peaks separated by lowlands, in other words any area surrounded by an environment not suitable for the island species.

In the 1960s, Robert MacArthur and E.O. Wilson began studying communities on islands to discover why large islands tend to have more species of a certain category (such as ferns, insects or birds) than do small islands. To explain these differences in species diversity, MacArthur and Wilson developed a general theory of island biogeography also called as **species equilibrium model** to identify the important determinants of species diversity on an island within a given set of physical characteristics. According to this model two factors will determine the number of species that eventually inhabit the island: (1) The rate at which new species immigrate to the island and (2) the rate at which species become extinct on the island. Immigration and extinction rates are, in turn, affected by two important features of the island: (1) Its size and (2) Its distance from the mainland, small islands will generally have lower immigration rates, because potential colonizers are less likely to reach a small island. For example, birds blown out to sea by a storm are more likely to land by chance on a larger island than on a small one. Small islands will also have higher extinction rates. They generally contain fewer resources and less diverse habitats for colonizing species to partition, increasing the likelihood of competition exclusion. Distance from the mainland is also important; for two islands of equal size, a closer island will have a higher immigration rate than the one farther away.

The immigration and extinction rates are also affected at any given time by the number of species already present on the island. As the number of species on the island increases, the immigration rate of new species decreases, because any individual reaching the island is less likely to represent a species that is not yet present. At the same time, as more species inhabit an island, extinction rates on the island increases because of the greater likelihood of competitive exclusion. These relationships are summarized in Figure 2.7, where immigration and extinction rates are plotted as a function of the number of species present on the island. The main prediction of this model is that eventually an equilibrium will be reached where the rate of species immigration matches the rate of species extinction (Figure 2.7a). The number of species at this equilibrium point is correlated with the islands size and the distance from the mainland. The Wilson/ MacArthur hypothesis of island biogeography predicts that the equilibrium species number is greater on a large island than a small island (Figure 2.7b), and greater on a near island than a far island (Figure 2.7c). Any equilibrium, of course, is dynamic; immigration and extinction continue, and the exact species composition may change over time.

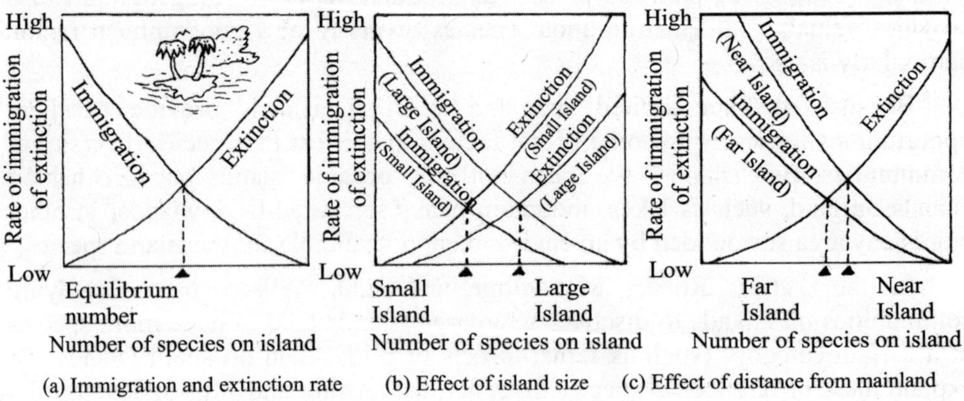

Figure 2.7. The theory of island biogeography. (a) The equilibrium number of species (triangle) on an island is determined by a balance between the immigration rate of new species and the extinction rate of species already on the island. (b) With time, large islands have a larger equilibrium number of species than smaller islands because of higher immigration rates and lower extinction rates on large islands. (c) Assuming equal extinction rates, an island near a mainland will have a larger equilibrium number of species than a more distant island because the immigration rate to a near island is higher than that to a more distant one.

Continental drift: The basis of biogeography

Evolution has dimension in space as well as time. The history of the earth helps explain the current geographical distribution of species. For example, the rise of volcanic islands such as the Galapagos creates new environments for founders that had reached there, and adaptive radiation fills many of the available niches with new species. On a global scale, the drifting of continents is the major geographical factor correlated with the spatial distribution of life.

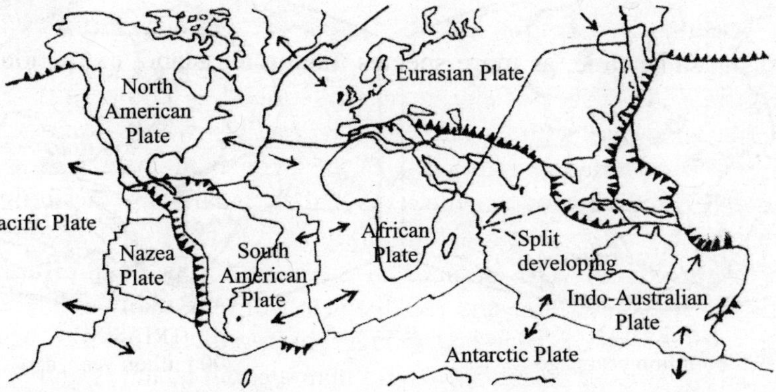

Figure 2.8 a: The modern continents on earth's crustal plates. The arrowheads indicate zones of tectonic events.

The continents are not fixed, but drift about earth's surface on great plates of crust floating on the hot, underlying mantle (Figure 2.8 a). Unless two land masses are embedded in the same plate, their positions relative to each other change. For example, North America and Europe are presently drifting apart at a rate of about 2 cm per year. Many important geological processes, including mountain building, volcanism, and earthquakes, occur at plate boundaries called as Subduction zone (Figure 2.8 b).

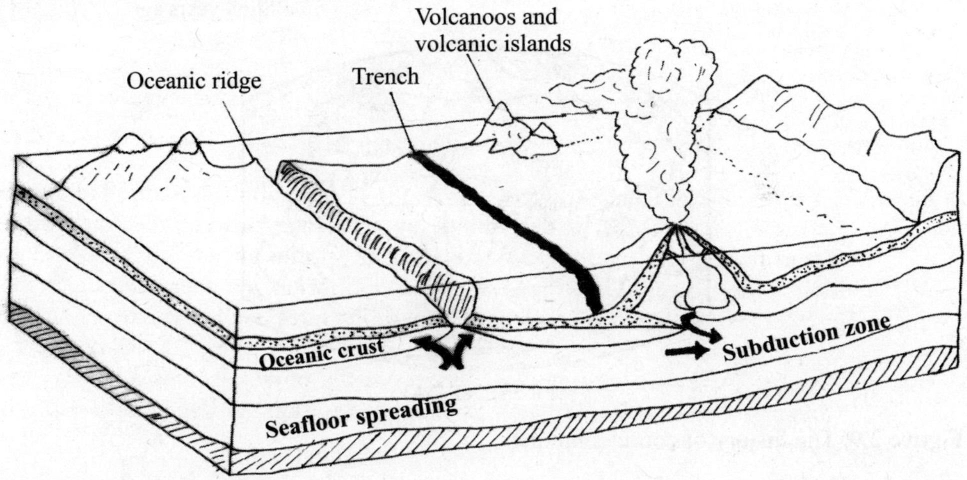

Figure 2.8b. At Subduction zone, where plates move towards each other, the denser plate dives below the less dense one, creating a trench. The abrasion at Sub-duction zones causes earthquakes, volcanic activities and mountain formation.

Two events related with continental drift had an especially strong influence on life. The first event about 250 million years ago, near the end of the Paleozoic era, plate movements brought all the land masses together to form a supercontinent that has been named as **Pangea** (Figure 2.9).

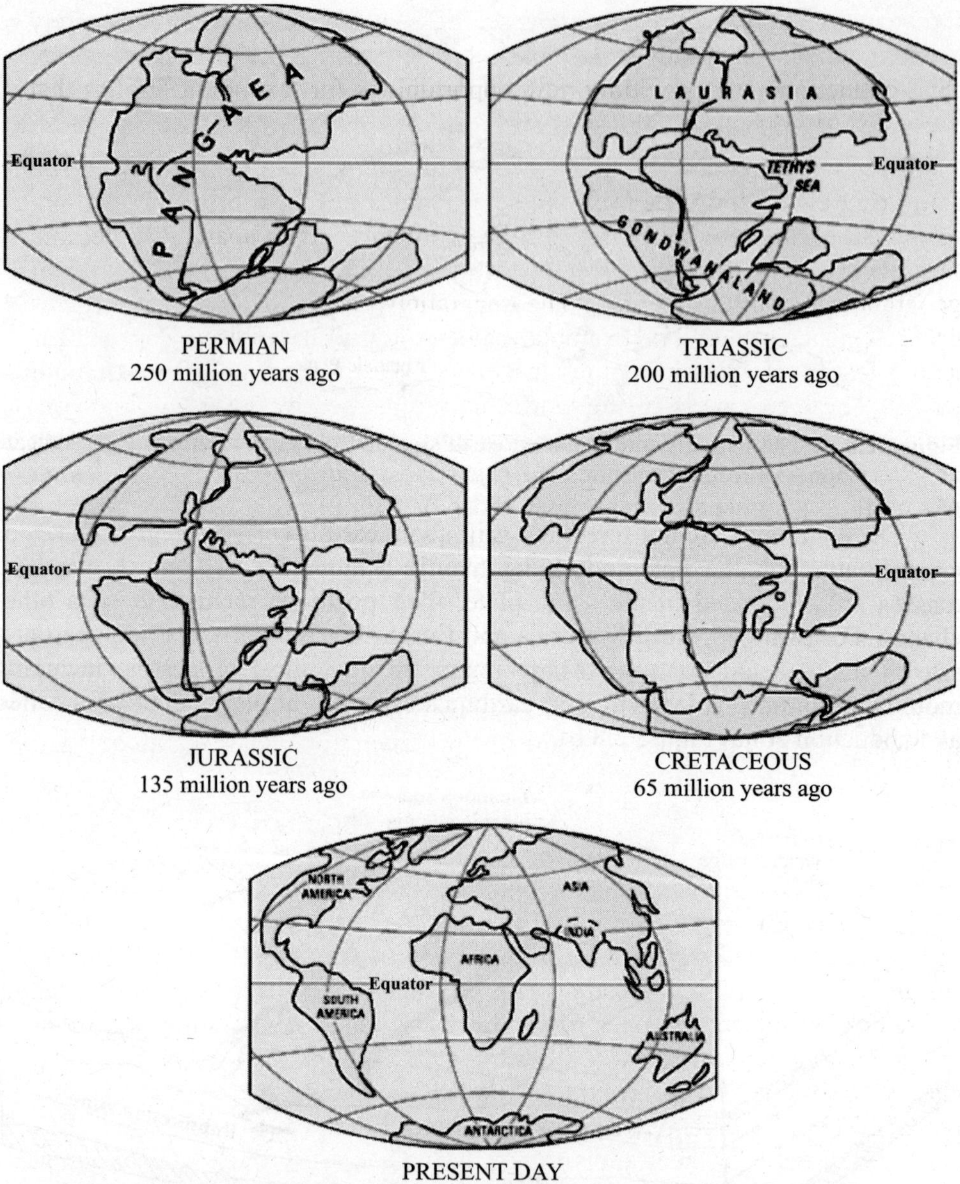

Figure 2.9. The history of continental drift

This event had a strong effect on life. Species that had been evolving in isolation came together and competed. When the land masses coalesced, the total amount of shoreline was reduced, and there is evidence that ocean basins increased in depth, which lowered the sea level and drained the shallow coastal seas. Most marine species inhabited shallow waters, and the formation of Pangea destroyed a considerable amount of that habitat. The continental interior, which has a drier and more erratic climate than the coastal regions, increased in area substantially when the land mass came together. Changing ocean currents also would have affected land life as well as sea life. The formation of Pangea surely had a tremendous environmental impact that reshaped biological diversity by causing extinctions and providing new opportunities for taxonomic groups that survived the crisis.

The second dramatic event in the history of continental drift was around 180 million years ago, during the Mesozoic era. Pangea began to break up, causing geographical isolation of land mass. As the continents drifted apart, each became a separate evolutionary arena, and the fauna and flora of the different biogeographical realms diverged. The separation pattern of continents solves many biological puzzles. For example, paleontologists have discovered similar matching fossils of Triassic reptiles in Ghana (West Africa) and Brazil (South America). These two parts of the world, now separated by 3000 km of ocean, were contiguous during the early Mesozoic era. Continental drift also explains much about the current distribution of organisms, such as why the Australian fauna and flora contrast so sharply with those of the rest of the world. The great diversity of marsupials, which fill the ecological roles in Australia analogous to those filled by placental mammals on other continents, is just one example of Australia's unique collection of species. Marsupials probably evolved first in what is now North America and reached Australia via South America and Antartica while the continents were still joined. The subsequent break-up of the southern continents set up Australia. On Australia, marsupials evolved and diversified, while on other continents placental mammals evolved and diversified.

Evidences from Fossils

Fossils are remains, traces or other direct evidences of past life forms. Most fossils form from burial of plants and animals in sediments; soft parts are more often consumed or decomposed but may leave imprints if buried rapidly. Most fossils are embedded in sedimentary rocks, weathered particles that provide strata from lower older layers to upper newer layers. Paleontologists study the fossil record based on boundaries between strata, where one mix of fossils gives way to another. Transitional links are intermediate between major groups. At any particular location, sedimentation is not continuous but occurs in intervals when the sea level changes or lakes and swamps dry up and refill. Even when a region is submerged, the rate of sedimentation and the types of sedimentary particles vary over time. As a result of these different periods of sedimentation, the rock forms in strata.

The organic substances of a dead organism buried in sediments usually decay rapidly. However, the hard parts that are rich in minerals, such as the shells of many invertebrates and protists and the bones and teeth of vertebrates, may remain as fossils.

Rarer than mineralized fossils are those that retain organic material. They are sometimes discovered as thin films pressed between layers of sandstone or shale. For example, palenontologists have discovered plant leaves millions of years old that are still green with chlorophyll and well enough preserved for their organic composition to be analyzed and the ultrastructure of their cells to be explored with an electron microscope. The most common fossilized plant material is pollen, which has a hard organic case that resists degradation.

Most fossils are not remnants of organisms but are found as rocks as replicas of the organisms. These fossils are formed when a dead organism locked in sediments decays and leaves an empty mould that becomes filled with minerals dissolved in water. The minerals may subsequently crystallize, forming a cast in the shape of the organism.

Some fossils are found as traces in the form of footprints, animal burrows or some animal activities. These fossils tell paleontologists about how the animals that left the trace fossils lived. For example, dinosaur tracks provide clues about the animal's locomotion.

In some cases entire animal bodies including the soft parts are fossilized – such fossils are formed only when there is no microbial decomposition. For example, an insect (Figure 2.10) got stuck in a drop of resin from a tree about 30 million years ago. The resin eventually hardened into amber. These different types of fossils help biologists to reconstruct the history of life.

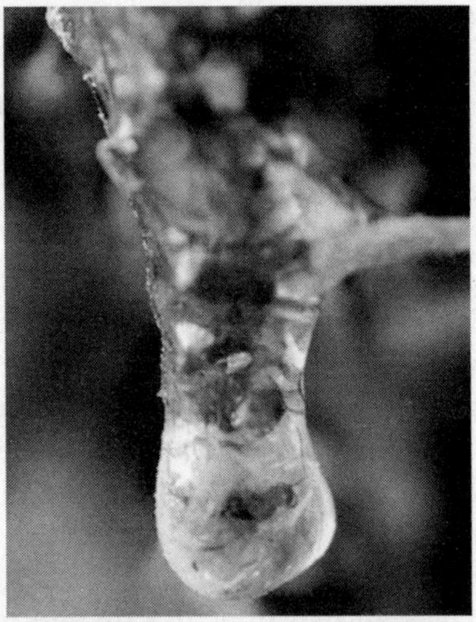

Figure 2.10. An insect trapped in amber

Isolation between populations encourages the evolutionary process, researchers rarely found transitional forms in the same place as the original forms. Thus, a complete evolutionary progression of fossils from most ancient to most modern has never been found in a single locality. A longstanding misperception of the fossil record of evolution is that fossil species form single lines of descent with unidirectional trends. Such a linear view of evolution is called **orthogenesis**, and has been rejected by paleontologists as a model of evolutionary change. The reality is much more complex than that, with numerous branching lines of descent and multiple morphologic trends (Figure 2.11).

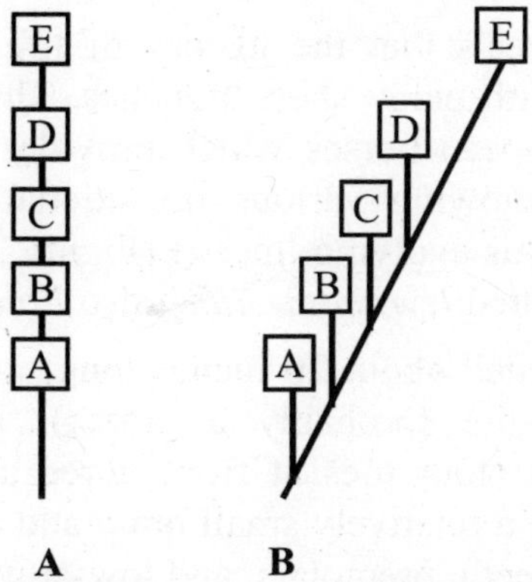

Figure 2.11. Comparison of orthogenesis with a branching phylogeny

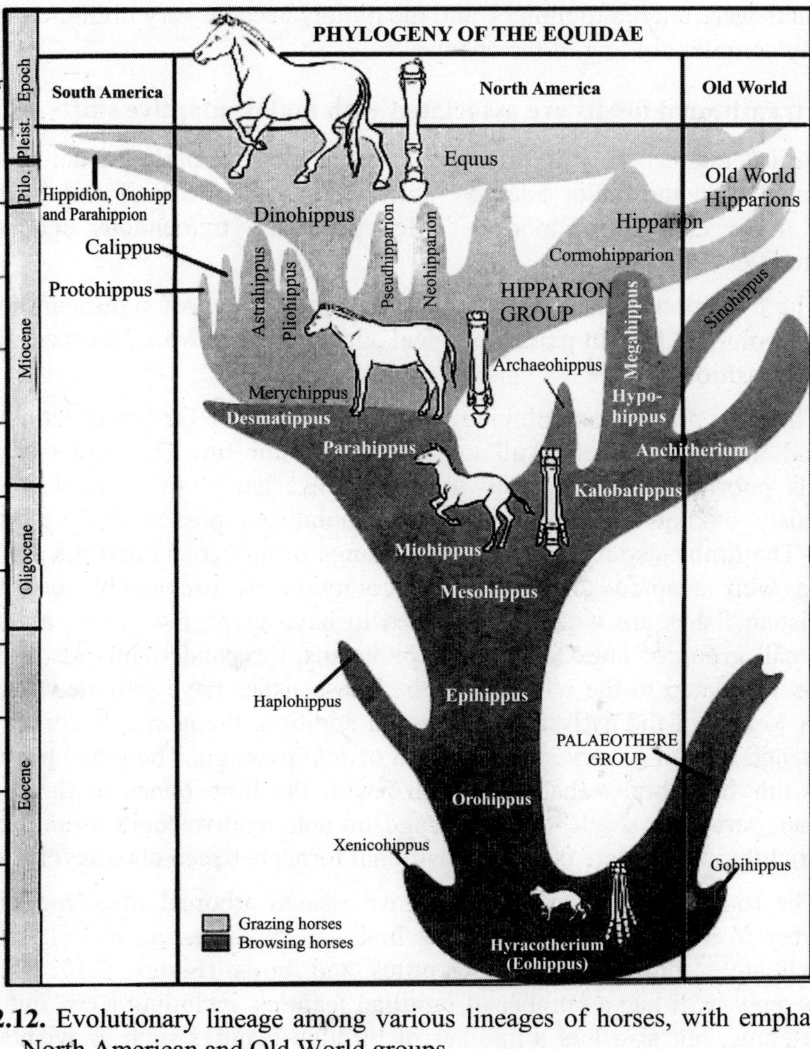

Figure 2.12. Evolutionary lineage among various lineages of horses, with emphasis on North American and Old World groups

The fossil record reveals that the history of life can be understood as a densely branching bush with many short branches. This is evident in fossils of North American and European horses which provides a stepwise evolutionary tree among vertebrates, showing various transitional stages and it does not represent a single continuous evolving lineage (Figure 2.12). The earliest known horse-like mammal was called *Hyracotherium* (also *Eohippus*).

Eohippus was very small about 20 inches long, weighing about 50 pounds and a generalized herbivore (probably a grazer). Besides the well-known difference in toe numbers (four toes at front, three at back), *Eohippus* had a narrow elongate skull with a relatively small brain and eyes forward in the skull. It possessed small canine teeth, premolars, and low-crowned simple molars. Over geologic time and within several lineages, the skull became much deeper, the eyes moved back, and the brain became larger. The incisors were widened, premolars were altered to molars, and the molars became very high-crowned with a highly complex folding of the enamel.

Some transitional fossils are associated with major adaptive shifts

Of special interest in the history of life are the morphological transitions associated with the major adaptive shifts from water to land and land to air. These major changes in mode of life opened up tremendous new adaptive opportunities for animal life.

The transition from water to land was one of the most significant events in animal evolution. Recent paleontological and systematic work has shed new light on this transition.

The most primitive amphibian yet known is the late Devonian *Icthyostega*, a tetrapod with a flattened skull and bearing a tail fin. The limbs were until recently poorly known, but new fossil evidence has come to light. The hand, previously unknown, shows that these amphibians possessed seven to eight digits. The limbs also had a very limited range of movement and the animal was not as well adapted to terrestrial locomotion as previously thought. The rhipidistian fishes are widely considered to have given rise to the amphibians. One small group of late Devonian rhipidistians, the panderichthyids, appears to be closely related to the ichthyostegids. These fishes have flattened skulls very similar to that of the early amphibians. In addition, the anal and dorsal fins are absent, and the tail is very similar to that of *Ichthyostega*. The lobed pectoral and pelvic fins have bones that homologize with the limb bones of the tetrapods. Whether part of a single direct lineage or not, ichthyostegid amphibians and panderichthyid fishes are clearly transitional forms between class level taxa.

The transition from land to air gave rise to arboreal life. One important discovery in 1861 of a true "missing link"- in this case, an animal that stood approximately midway between reptiles and birds (Figure 2.13) known as *Archaeopteryx*. It had a number of reptilian features, including teeth and a tail of 21 vertebrae, but also had a number of birdlike features such as wishbone and feathers. It was considered as a probable cousin to the lineage running from reptiles (dinosaurs) to birds.

Figure 2.13. (a) The fossil Archeopteryx found in the Upper Jurassic limestones of Bavaria (b) Restoration of the bird's appearance

Methods used by paleontologists to date fossils

To trace evidences in favour of evolution fossils are reliable historical data only if we can determine their ages. Here two methods are discussed for determining the age of fossils and where fossils fit on the geological time-scale.

1. Relative dating

Relative dating determines the relative order of the past events, without necessarily determining the absolute age. The trapping of the dead organisms in sediments freezes fossils in time. Thus the fossils in each stratum of the sedimentary rock are a local sample of the organisms that existed at the time that sediment was deposited. Because younger sediments are superimposed upon older ones, it tells the relative age of fossils.

Relative dating by biostratigraphy is the preferred method in paleontology. Biostratigraphy is the branch of stratigraphy which focuses on correlating and assigning relative ages of rock strata by using the fossil assemblages contained within them. Usually the aim is correlation, demonstrating that a particular horizon in one geological section represents the same period of time as another horizon at some other section. The fossils are useful because sediments of the same age can look completely different because of local variations in the sedimentary environment. For example one section might have been made up of clays and marls while another has more chalky limestones, but if the fossil species recorded are similar, the two sediments are likely to have been laid down at the same time.

The strata at one location can be correlated with strata at another location by the presence of similar fossils known as **index fossils**. The best index fossils for correlating strata that are far apart are the shells of sea animals that were widespread. To work well, the fossils used must be widespread geographically, so that they can occur in many different places. They must also be short lived as a species, so that the period of time during which they could be incorporated in the sediment is relatively narrow. The longer the species lived, the poorer the stratigraphic precision, so fossils that evolve rapidly are favoured over forms that evolve much more slowly.

2. Absolute dating

In absolute dating the age is given in years instead of relative terms such as before and after, early and late. Radiometric dating is the method most often used to determine the age of rocks and fossils on a scale of absolute time. Fossils contain isotopes of elements that accumulated in the organisms when they were alive. Because each radioactive isotope has a fixed rate of decay, it can be used to date a specimen. An isotope's half-life, the number of years it takes for 50% of the original sample to decay, is unaffected by temperature, pressure, and other environmental variables.

3. Radiocarbon Dating

One of the most widely used and well-known absolute dating techniques is carbon-14 (radioisotope) dating, which is used to date organic remains. This is a radiometric technique since it is based on radioactive decay. Carbon-14 is an unstable isotope of normal carbon, carbon-12, the half-life of carbon-14 is 5,730 years. Cosmic radiation entering the earth's atmosphere produces Carbon-14, while an organism is alive it assimilates the different isotopes of each element in proportions determined by their relative abundances in the environment. Carbon 14 is taken in trace quantities, along with much larger quantities of the more common carbon 12. After an organism dies, it is covered with sediment, and eventually becomes consolidated into a layer of rocks as the sediment is compressed. From the time the organism dies and ceases to assimilate carbon, the amount of carbon 14 relative to carbon 12 in the fossil declines due to radioactive decay. The age can be determined by the ratio of the two isotopes to learn how much half-life reductions have occurred since it died. For example, if the ratio of carbon 14 to carbon 12 in the fossil was found to be one-fourth that of a living organism, this fossil would be about 11,460 years old.

Because the half-life of carbon 14 is 5,730 years, carbon dating is only reliable about up to 50,000 years. To date older fossils paleontologists use radioactive isotopes with longer half-lives. For example, paleontologists use potassium 40 whose half-life is 1.3 billion years and uranium 238, which has a half-life of 4.5 billion years, has been used as a radiometric clock to date fossil-containing rocks of the Cambrian period.

These radioactive isotopes used in radiometric techniques are also called as **clocks of the rocks**.

Geological time-scale

Geological time scale is the tabular presentation of sequence and duration of different eras and periods with their dominant form or life. It has five principle eras: **Proterozoic, Archoeozoic, Paleozoic, Mesozoic** and **Cenozoic.** Geological history of the earth is divided into eras, periods, and epochs (Figure 2.14). Fossil record provides relative dating of rock layers; top layers of rock are younger than the lower layers. Absolute dating method uses radioactive isotopes. Isotopes each have particular half-life or time it takes for half of isotope to decay and become non-radioactive. Carbon-14 (14C) used to date organic matter; half decays to 14N each 5,730 years; limited to about last 50,000 years. Half of potassium-40 (40K) decays to argon-40 (40Ar) each 1.3 million years; estimates age of younger rocks. Uranium-238 decays to lead-207; estimates age of older rocks.

Figure 2.14. Geological Time scale with Dates in Millions of year (in parenthesis)

Era	Period	Epoch	Aquatic Life	Terrestrial Life
Cenozoic (65) The Age of Mammals	Quaternary (1.8)	Recent (0.1)		Humans in the new world
		Pleistocene (1.8)	Periodic glaciation Continental drift continues	First humans
	Tertiary (65)	Pliocene		Hominids and pongids
		Miocene	All modern groups present	Monkeys and ancestors of apes
		Oligocene		Adaptive radiation of birds
		Eocene		Adaptive radiation of mammals
		Paleocene		
Mesozoic (251) The Age of Reptiles	Cretaceous (146)	Still attached: N. America & N. Europe; Australia & Antarctica	Modern bony fishes, Extinction of ammonites, plesiosaurs, ichthyosaurs	Extinction of dinosaurs, pterosaurs. Rise of woody angiosperms, snakes; first placental mammals (Eutheria)
	Jurassic (200)	Africa & S. America begin to drift apart	Plesiosaurs, ichthyosaurs abundant; first diatoms Ammonites again abundant Skates, rays, and bony fishes abundant	Dinosaurs dominant; first angiosperms First mammals; Archaeopteryx; first lizards Adaptive radiation of dinosaurs; insects abundant
	Triassic (251)	At the end Pangaea splits into Laurasia and Gondwana	First plesiosaurs, ichthyosaurs. Ammonites abundant at first. Rise of bony fishes	Adaptive radiation of reptiles: thecodonts, therapsids, turtles, crocodiles, first dinosaurs
Paleozoic (542)	Permian (299)	Appalachian Mts. formed; periodic glaciation and arid climate	Extinction of trilobites, placoderms	Reptiles abundant: cotylosaurs, pelycosaurs. Cycads, conifers, ginkgos

Era	Period	Epoch	Aquatic Life	Terrestrial Life
	Carboni-ferous **Periodic aridity**	Pennsylvanian (320)	Ammonites, bony fishes	First reptiles Coal swamps
		Mississippian (359)	Adaptive radiation of sharks	Forests of lycopsids, sphenopsids, and seed ferns Amphibians abundant Land snails
	Devonian (416) The "Age of Fishes"	Extensive inland seas	Placoderms, cartilaginous and bony fishes. Ammonites, nautiloids	Ferns, lycopsids, and sphenopsids First gymnosperms
			Adaptive radiation of ostracoderms, erypterids	First amphibians Arachnids (scorpions)
	Silurian (444)	Mild climate; inland seas	Nautiloids, Pilina, other mollusks	First insects
	Ordovician (488)	Mild climate, inland seas	Trilobites abundant First jawless vertebrates	First fungi First plants (liverworts?) First millipedes?
	Cambrian (542) Periodic glaciation		Trilobites dominant. Euryp-terids, crustaceans mollusks, echino-derms sponges, cnidarians, anne-lids, and tunicates present	No fossils of eukaryotes, but phylogenetic trees suggest that lichens, mosses, perhaps even vascular plants were present.
Proterozoic (2500)	Ediacaran (620)		Fossils rare but evidence of several types of invertebrates	No fossils of eukary-otes, but phylogene-tic trees suggest that lichens, mosses, per-haps even vascular plants were present towards the end.

Evidences from comparative anatomy

There are many forms of evidence for **evolution**. One of the strongest forms of evidence is comparative **anatomy**; comparing structural similarities of organisms to determine their evolutionary relationships. Organisms with similar anatomical features are assumed to be relatively closely related evolutionarily, and they are assumed to share a common ancestor. As a result of the study of evolutionary relationships, anatomical similarities and differences are important factors in determining and establishing classification of organisms.

Some organisms have anatomical structures that are very similar in embryological development and form, but very different in function. Similarity in characteristics resulting from common ancestry is known as **homology,** and such anatomical signs of evolution are called as **homologous structures**. Since these structures are so similar, they indicate an evolutionary relationship and a common ancestor of the **species** that possess them. A clear example of homologous structures is the forelimb of **mammals**. When examined closely, the forelimbs of humans, whales, cats, and bats all are very similar in structure (Figure 2.15). Each possesses the same number of bones, arranged in almost the same way. While they have different external features and they function in different ways, the embryological development and anatomical similarities in form are striking. By comparing the anatomy of these organisms, scientists have determined that they share a common evolutionary ancestor and in an evolutionary sense, they are relatively closely related. The evolution of structurally similar and functionally dissimilar parts of organisms in different directions is called divergent evolution. It indicates the adaptation of structures in different adaptive zones.

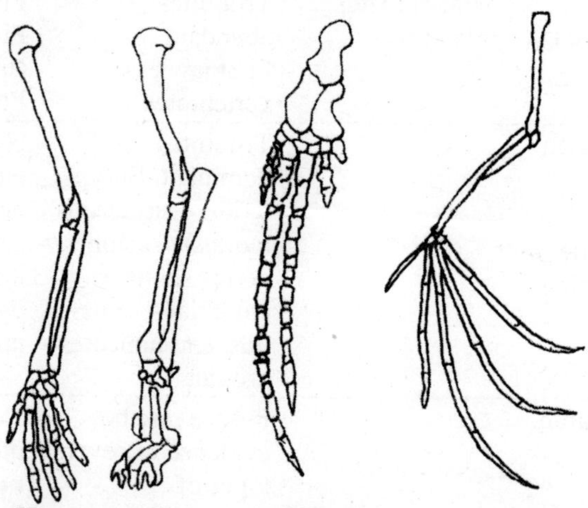

Figure 2.15. Homologous structures, forelimbs of mammals.

Other organisms have anatomical structures that function in very similar ways, however, morphologically and developmentally these structures are very different. These are called **analogous structures**. Since these structures are so different even though they have the same function, they do not indicate an evolutionary relationship nor that two species share a common ancestor. For example, the wings of a bird and dragonfly both serve the same function; they help the **organism** to fly. However, when comparing the anatomy of these wings, they are very different. Bird wings have bones inside and is covered with feathers, while the dragonfly wing is missing both of these structures. They are analogous structures. Thus, by comparing the anatomy of these organisms, scientists have determined that **birds** and **dragonflies** do not share a common evolutionary ancestor, nor that, in an evolutionary sense, they are closely related. Analogous structures are evidence that these organisms evolved along separate lines. The evolution of different organisms, or parts of organisms, in such similar directions is called **convergent evolution,** indicating that selection for similar habitats in different evolutionary lineages could occasionally lead to functionally similar anatomical structures.

Some of the most interesting homologous structures are the **vestigial organs.** These structures are anatomical features that are still present in an organism (although often reduced in size) even though they no longer serve a function. When comparing anatomy of two organisms, presence of a structure in one and a related, although vestigial structure in the other is evidence that the organisms share a common evolutionary ancestor and that, in an evolutionary sense, they are relatively closely related. Whales, which evolved from land mammals, have vestigial hind leg bones in their bodies (Figure 2.16). While they no longer use these bones in their marine **habitat**, they do indicate that whales share an evolutionary relationship with land mammals. Humans have more than 100 vestigial structures in their bodies; some of the vestigial structure in humans is shown in Figure 2.17.

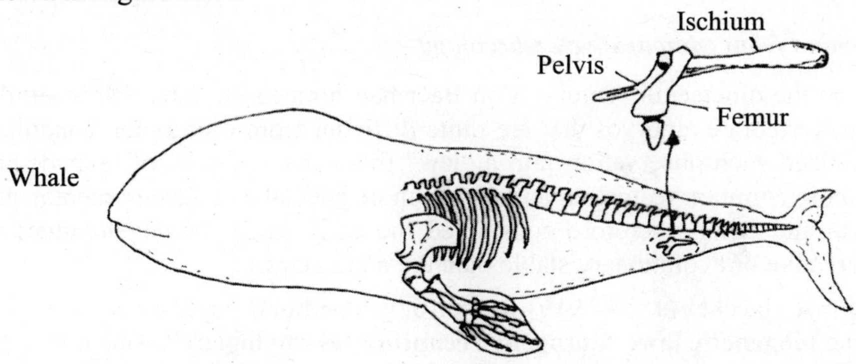

Figure 2.16. Hind limb of whale.

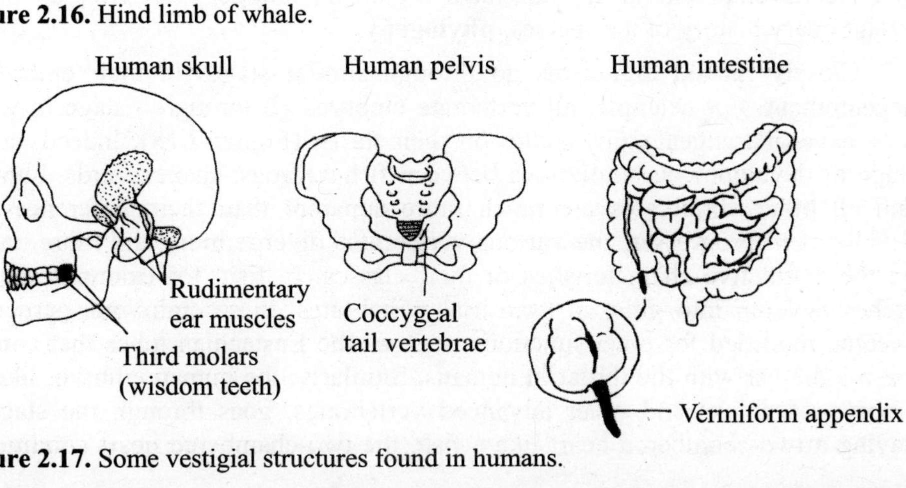

Figure 2.17. Some vestigial structures found in humans.

On a superficial level, vestigial organs may seem to support Lamarck's use and disuse concept, but now it is well known that the effects of an individual's use of body structure are not inherited by that individual's offspring. On the other hand, vestigial organs are evidence of evolution by natural selection. Because it would be wasteful to continue providing blood, nutrients, and space to an organ that no longer has a major function, natural selection would tend to favour individuals with reduced version of those organs, and thereby tend to phase out obsolete structures. Ultimately, structural changes such as the adaptation of the tail as the main propulsive structure and the reduction of the hindlimbs in whales involve changes in patterns of gene expression during embryonic development, because the embryonic processes affect the functioning of the adult organism, they are themselves subjected to natural selection. Thus, vestigial organs represent the changes in an organism's embryonic development by natural selection.

Comparative anatomy is an important tool that helps determine evolutionary relationships between organisms and whether or not they share common ancestors. However, it is also important evidence for evolution. Anatomical similarities between organisms support the idea that these organisms evolved from a common ancestor. Thus, the fact that all **vertebrates** have four limbs and gill pouches at some part of their development indicates that evolutionary changes have occurred over time resulting in the diversity we have today.

Evidences from comparative embryology

Early in the nineteenth century, Von Baer had noticed the remarkable similarity among vertebrate embryos that are quite different from each other as adults. He generalized such observation into a "law" that early embryos of related species bear more common features than later do more specialized developmental stages. Darwin and others therefore considered the early stages of development more conservative or evolutionary stable than the adult stages.

Ernst Haeckel (1834-1919) further developed and popularized this concept into the **biogenetic law: "ontogeny recapitulates phylogeny".** This notion holds that the development of an individual organism, ontogeny, is a replay of the evolutionary history of the species, **phylogeny**.

Closely related organisms go through similar stages in their embryonic development. For example, all vertebrate embryos go through a stage in which they have pharyngeal (gill) arches on their throat (Figure 2.18). Indeed, at this stage of development, similarities between fishes, frogs, snakes, birds, humans, and all other vertebrates are much more apparent than their differences. As development progresses, the various vertebrates diverge more and more, taking on the distinctive characteristics of their classes. In fish, for example, the gill arches develop into gills; in terrestrial vertebrates, these embryonic structures become modified for other functions, such as the Eustachian tubes that connect the middle ear with the throat in humans. Similarly, the human embryo, like the embryo of chicks and other advanced vertebrates, goes through the stage of having a two-chambered heart like a fish, the two-chambered heart condition is restored in adult fishes, later on the heart becomes three-chambered and this condition is restored in almost all amphibians and most reptiles. In mammals and birds the heart becomes four-chambered and this condition is restored in their adult forms.

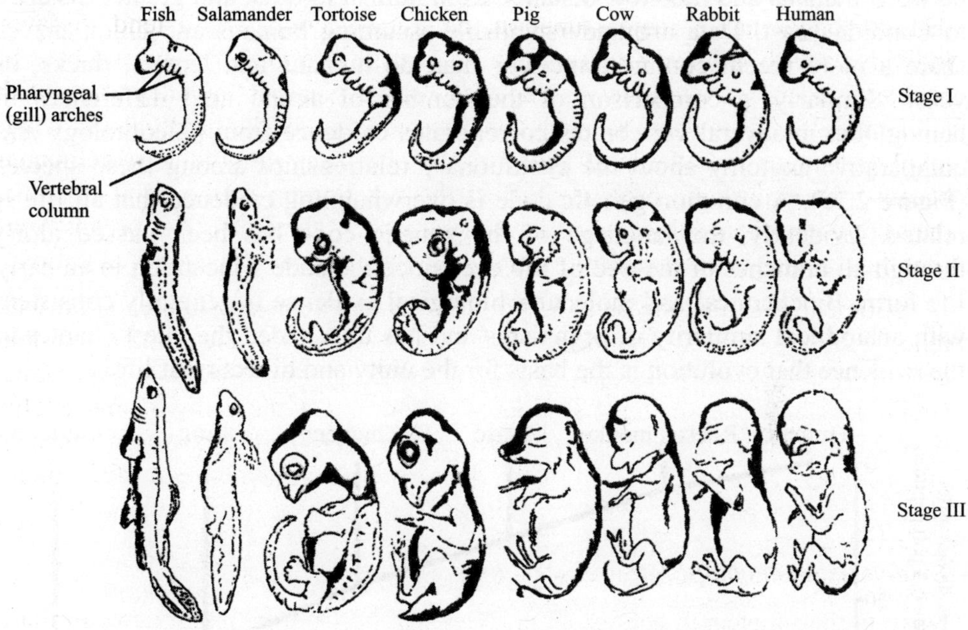

Figure 2.18. Different vertebrate embryos at different stages of development. Note that each of the embryos begins with a similar number of pharyngeal (gill) arches (pouches below the head) and a similar vertebral column. In later stages of development, these and other structures are modifies to yield each vertebrate group's characteristic form.

Moreover, development of all **triploblastic** animals starts from zygote and undergoes similar changes to form gastrula having three primary germ layers (ectoderm, mesoderm and endoderm) which have the same fate in organogenesis. Earlier embryos of different vertebrates resemble in possessing similar structures like gills, slits, notochord, tail, etc. Not only this, in the course of development at different stages an embryo looks like the embryo of different phyla from which the given organism has evolved.

Evidences from Biochemistry and Molecular Biology

Almost all living organisms use the same basic biochemicals: DNA, ATP, many identical enzymes, DNA triplet code, 20 amino acids, introns, and hypervariable regions. Prokaryotes i.e. true bacteria (Domain Eukarya), do not have introns. This points to a long period of time since all living things shared common ancestry. Similarity of biochemistry is explained by descent from common ancestor.

DNA base sequences differences in DNA between a number of organisms shows less difference the more closely related they are; for example, 2.5% difference between humans and chimpanzees but 42% difference between humans and lemurs. Amino acid sequences of cytochrome c show similarity between humans and monkey, distance from human to duck and greater distance to Candida yeast. Data are understandable assuming humans and chimpanzees share a more recent common ancestor than do humans and lemurs, ducks, or yeast. Similarly, a comparison of the number of amino acid differences in hemoglobin in several vertebrates corroborates evidence from paleontology and comparative anatomy about the evolutionary relationships among these species (Figure 2.19). A common genetic code is overwhelming evidence that all life is related. Evidently, the language of the genetic code has been passed along through all branches of the tree of life ever since the code's inception in an early life form. Biochemical and molecular biological evidence is generally consistent with anatomical similarity of organisms and has thus added the latest chapter to the evidence that evolution is the basis for the unity and diversity of life.

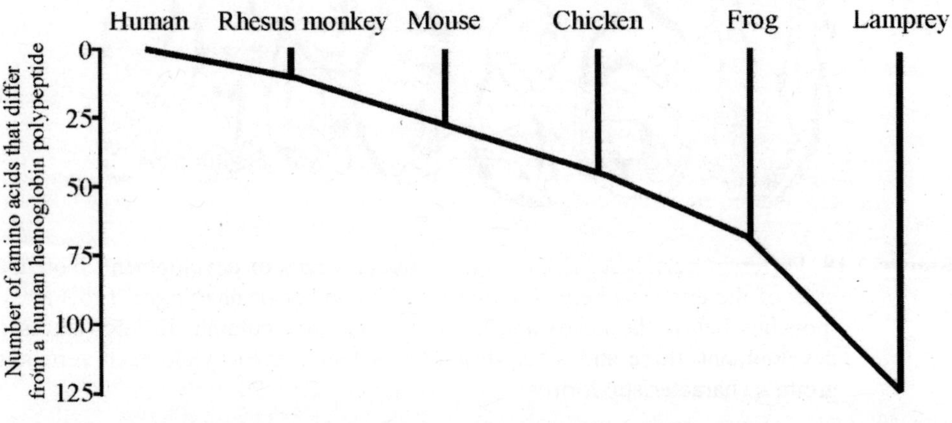

Figure 2.19. Molecular data and the evolutionary relationships of vertebrates. Branch points in this vertebrate tree are defined by the number of amino acids that differ from a polypeptide of human hemoglobin. The molecular data are compatible with vertebrate relationships deduced from the fossil record and comparative anatomy

■■■■

Chapter 3

THE MECHANISM OF EVOLUTION

3.1 The foundation of genetics
- 3.1.1 Mendel's Experiments and the laws of inheritence
- 3.1.2 Monohybrid cross
- 3.1.3 Dihybrid cross
- 3.1.4 Some Exception to Mendal's Laws

3.2 Population genetics
- 3.2.1 Integration of Darwinian selection and Mendelian inheritance to give rise modern evolutionary theory
- 3.2.2 Genetic structure of a population
- 3.2.3 Hardy-Weinberg theorem for non evolving populations
- 3.2.4 Hardy-Weinberg equilibrium for two loci
- 3.2.5 Hardy-Weinberg equilibrium for multi loci
- 3.2.6 Conditions required to maintain Hardy Weinberg equilibrium

3.3 Microevolution and its causes
- 3.3.1 Genetic drift
- 3.3.2 Gene flow
- 3.3.3 Mutation
- 3.3.4 Non random mating
- 3.3.5 Natural selection

3.4 Macroevolution
- 3.4.1 Mammal's evolution from reptilian ancestors: A closer look

3.5 Extinction and Radiations
- 3.5.1 Cambrian explosion: One of the most important adaptive radiations in the history of life.
- 3.5.2 Mass extinctions
  - 3.5.2.1 Permian extinction
  - 3.5.2.2 Extinction of dinosaurs (The Cretaceous extinction)
  - 3.5.2.3 Sixth mass extinction

3.6 Coevolution
- 3.6.1 Insect-plant coevolution

3.7 Adaptation and Natural selection
- 3.7.1 Adaptation

3.7.2 Types of adaptation
3.7.3 Coadaptation
3.8 Natural selection as the mechanism of adaptive evolution
3.8.1 Concept of fitness
3.8.2 The effect of selection on a varying characteristic can be stabilizing, directional, or diversifying

3.1 THE FOUNDATION OF GENETICS

Five thousand years ago or earlier, people were using applied genetics in the form of plant and animal breeding. The foundation for the science of genetics was laid in 1866, when Gregor Mendel used varieties of peas to conduct experiments on inheritance. Mendel's research was ignored until the turn of the twentieth century. Plant breeders showed that both parents contribute equally to inheritance. Plants have some desirable characteristics for genetic studies. Plants can be grown in large quantities. They produce a large numbers of offspring. Many have both male and female reproductive organs in the same plant, which allows self-fertilization. The generation time is relatively short. It is easy to control which individuals should mate.

Josef Gottlieb Kölreuter made a few observations that Mendel later found useful. Kölreuter's research focused on the relative contribution of males and females to the genetics of plants. Reciprocal crosses helped prove that both male and female parents contribute equally to the characteristics inherited by their offspring. An example is a cross of a male white plant with a female pink plant, and the reciprocal cross of a male pink with a female white plant. The resulting progeny have the same appearance.

Before the acceptance of Mendel's research, the concept of blending was favored. An example of this form of inheritance would be purple progeny resulting from red and blue parents. As a consequence of blending, the purple color would be forever blended and could not be separated. Mendel's discoveries were overlooked for decades.

Gregor Mendel was a monk with scientific training in mathematics, physics, and biology. Mendel presented his nine-year-long project orally in 1865 and in writing in 1866. His data challenged the blending concept, but he was ahead of his time. His theory was ignored, perhaps because his biological peers were not accustomed to reviewing mathematical data. Even Darwin, whose evolutionary theory rests on genetic variation among individuals, failed to understand Mendel’s point and relied on the blending concept.

In 1900, Hugo de Vries, Karl Correns, and Erich von Tschermak each independently published papers on the quantitative outcomes from crosses. Each cited Mendel's then rediscovered 1866 paper. By 1900, meiosis had been described and Mendel’s ideas were finally accepted.

3.1.1 Mendel's Experiments and the laws of inheritance

Mendel devised a careful research plan. He chose garden peas as his subjects as they are easily grown and their pollination is easily controlled. He controlled pollination by manually moving pollen between plants. He also allowed the plants to self-pollinate (using emasculation and bagging). Mendel examined varieties of peas for heritable characters and traits for his study. A character is a feature, such as flower color. A trait is a particular form of a character, such as white flowers. "Heritable" means the trait can be passed from parent to progeny. Mendel looked for characters that had well-defined alternative traits and that were true breeding. For his experiments he selected seven contrasting characters: plant height, flower position, pod shape, pod color, seed shape, seed color, and flower color (Figure 3.1). A trait is true breeding when it is the only trait that occurs through many generations of breeding individuals that share that trait. A true-breeding white-flowered plant would have only white-flowered progeny when crossed with others in its strain.

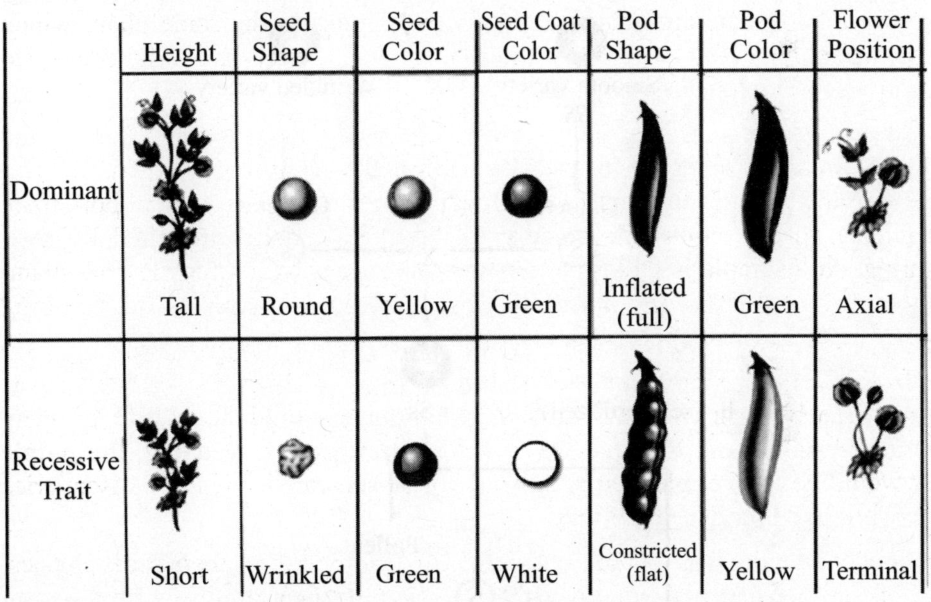

Figure 3.1. Seven contrasting characters selected by Mendel from garden pea.

True-breeding plants when used for crossing with other plants that have an alternative trait, are called the parental generation and designated *P*. The progeny from the cross of the *P* parents are called the first filial generation, designated F_1. When F_1 individuals are crossed with each other or self-fertilized, their progeny are designated F_2.

Mendel's well-organized plan allowed him to observe and record the traits of each generation in sufficient quantity to explain through analysis the relative proportions of the kinds of progeny. His paper is recognized today as a model of clarity.

3.1.2 Monohybrid cross

Mendel's Experiment 1 examined a monohybrid cross. He crossed true-breeding plants that differed for a given character. A monohybrid cross involves one (mono) character and different (hybrid) traits. Pollen from true-breeding pea plants with wrinkled seeds (one trait) was placed on stigmas of true-breeding plants with spherical/smooth seeds (another trait).

The F_1 seeds were all spherical; the wrinkled trait failed to appear at all. Because the spherical trait completely masks the wrinkled trait when true-breeding plants are crossed, the spherical trait is called *dominant*, and the wrinkled trait is called *recessive*.

The F_1 plants were allowed to self-pollinate. This step was the monohybrid cross. *This is also called an F_1 cross. Self-pollination is sometimes called* ***selfing.***

The progeny, called F_2, were examined: 5,474 were spherical and 1,850 were wrinkled in ratio 3:1 (Figure 3.2).

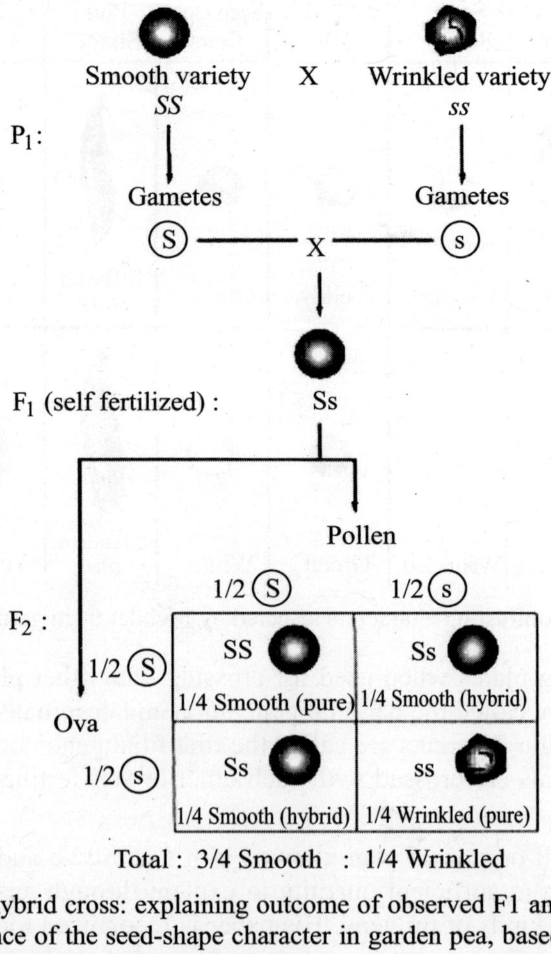

Figure 3.2. Monohybrid cross: explaining outcome of observed F1 and F2 results for the inheritance of the seed-shape character in garden pea, based on segregation of alleles *S* (dominant) and *s* (recessive).

Mendel proposed that the units responsible for inheritance were discrete particles. They existed within an organism in pairs, separated during gamete formation, and retained their integrity. This is called the **Particulate theory**, which is in sharp contrast to the Blending theory. Each pea has two units of inheritance for each character. During production of gametes, only one of the pair members for a given character passes to the gamete. When fertilization occurs, the zygote gets one from each parent, restoring the pair.

Mendel's units of inheritance are now called **genes.** Different forms of a gene are called **alleles**. Each allele is given a symbol. In the case of wrinkled seeds, *S* might represent smooth and *s* wrinkled. By convention, uppercase *S* represents the dominant; lowercase *s* represents the recessive.

True-breeding individuals would have two copies of the same allele. Wrinkled would be *ss* (two copies of same allele = *homozygous*). Smooth true-breeding would be *SS* (two copies of same allele = *homozygous*). Some smooth-seeded plants could be *Ss*, although they would *not* be true-breeding. Individuals that are smooth, but had a wrinkled parent, are *heterozygous*: *Ss*.

When an organism is studied for three different genes and has the alleles *AABbCC*, it is homozygous for *A* and *C* genes but heterozygous for the *B* gene. The physical appearance of an organism is its *phenotype*. Wrinkled-seed would be a phenotype. The actual composition of the organism's alleles for a gene is its *genotype*: *Ss* is a genotype.

Organisms have many different genes — some have thousands, and complex organisms have 10 times that number. Most of these genes are yet to be described in terms of the DNA sequence or the amino acid sequence of the gene product.

Mendel's First law says that alleles segregate. When an individual produces gametes, alleles separate, so each gamete receives one member of the pair of alleles. **This is Mendel's First law, the Law of segregation** (Figure 3.3).

When fertilization occurs, pairs are re-established by receiving one copy from each parent. The Punnett square is a simple box-like device that helps us to consider all genetic combinations and can provide clarity by showing the expected frequencies of genotypes. The *S* and *s* symbols represent the single allele each gamete receives. Fertilization provides the two alleles for the new individual, one from the male (sperm) and one from the female (egg).

The Punnett square shows that the genotypes and associated ratios for a monohybrid cross are **1 *SS*: 2 *Ss* :1 *ss*.** Any progeny with an *S* would have the dominant (smooth) phenotype, so the phenotypic ratio is 3 smooth to 1 wrinkled.

Now it is known that a gene is a portion of the chromosomal DNA that resides at a particular site called a *locus* (plural is *loci*). The gene codes for a particular function. Mendel arrived at the Law of segregation with no knowledge of meiosis or chromosomes. The mechanism of chromosome separation in **meiosis I** today explains his Law of segregation. Mendel verified his hypothesis by performing a test cross.

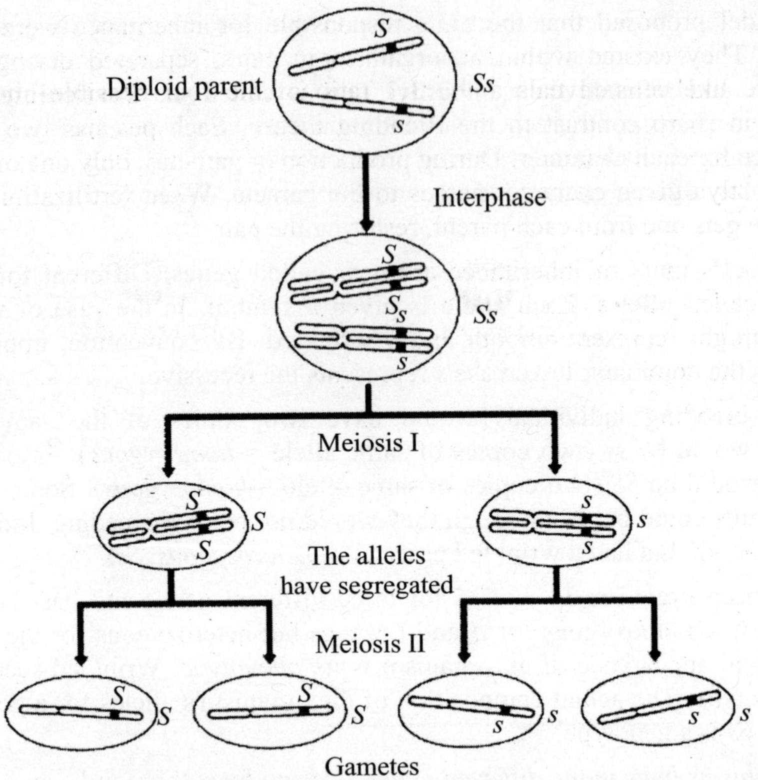

Figure 3.3. Explanation for segregation of gametes according to the Law of segregation.

A test cross can determine the genotype (*heterozygous* or *homozygous*) of an individual with a dominant trait. It involves crossing the individual to a true-breeding recessive (homozygous recessive). If the unknown is heterozygous, approximately half the progeny will have the dominant trait and half, the recessive trait **(Ratio 1:1)**. If the unknown is homozygous dominant, all the progeny will have the dominant trait.

Back cross is a cross of F_1 with any one of the parent. Generally back cross with homozygous recessive parents is made in plant breeding application to ensure maximum homozygosity in order to obtain pure line variety.

3.1.3 Dihybrid cross

Mendel's Second law says that alleles of different genes assort independently. It describes the outcome of **dihybrid** (two characters) crosses, or hybrid crosses involving additional characters. A dihybrid is an individual that is a double heterozygote (e.g., with the genotype *SsYy*) (Figure 3.4).

As mentioned, Mendel's Second law states that the *Ss* alleles assort into gametes independently of the *Yy* alleles. The dihybrid, *SsYy*, produces gametes that have one allele of each gene. Four different gametes are possible and will be produced in equal proportions: *SY, Sy, sY,* and *sy*. Random fertilization of gametes yields the outcome visible in the Punnett square. Note its 4 × 4 table construction to accommodate 16 possible phenotypes. Filling in the table and adding the like cells reveals a **9:3:3:1** ratio of the four possible phenotypes (smooth yellow, smooth green, wrinkled yellow, and wrinkled green). **The Law of independent assortment** states that alleles of different genes assort independently of one another during gamete formation (Figure 3.5).

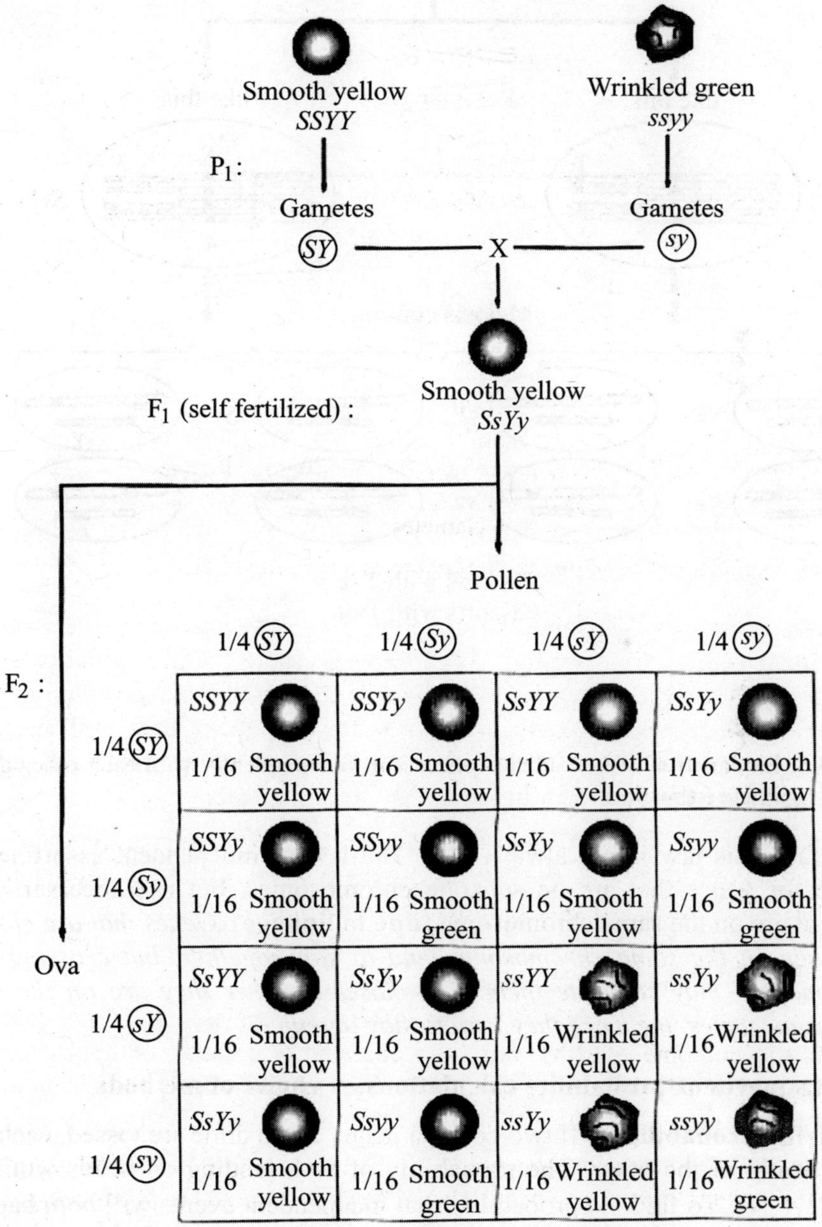

Figure 3.4. Results of Mendelian dihybrid cross

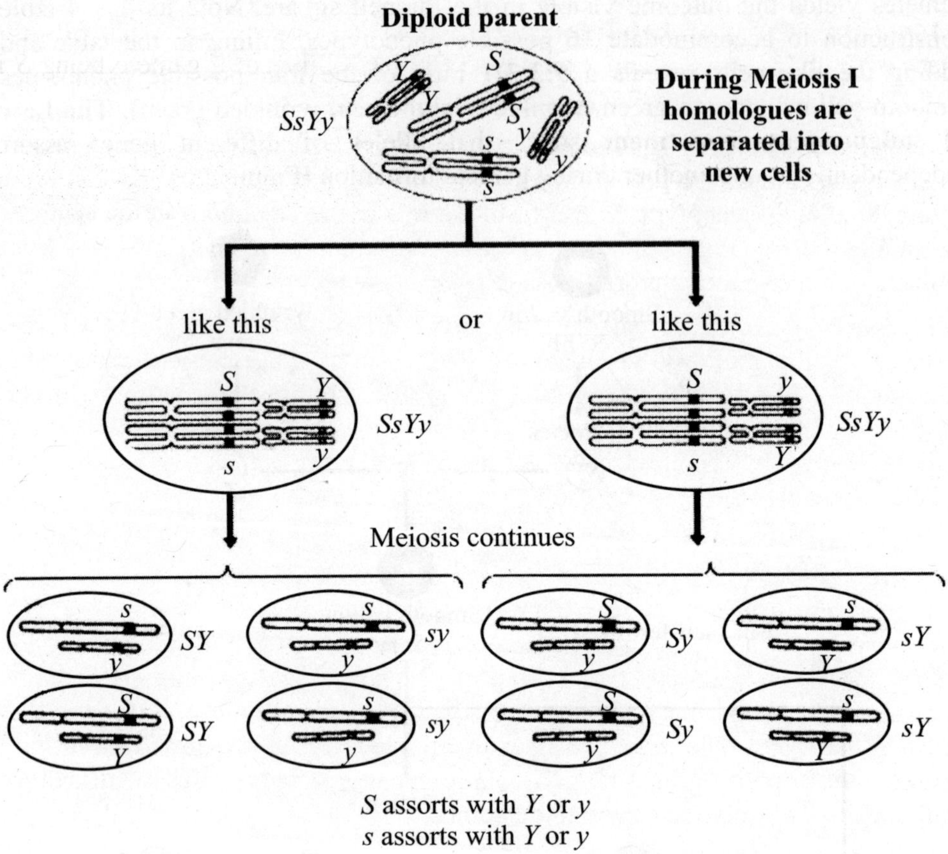

Figure 3.5. Explanation of how alleles pair assort independently with each other during gamete formation

In fact, this law is not always true. The law of independent assortment is accurate for genes that are on separate chromosomes, but not necessarily for genes that are on the same chromosome (**due to linkage**). *Genes that are close to each other on the same chromosome tend to stay together, but crossing over during meiosis may separate them. The closer together they are on the same chromosome genes, the more they tend to stay together.*

Punnett squares or probability calculations: A choice of methods

Multiplying probabilities: If two coins, a penny and a dime are tossed, each acts independently of the other. The probability of both landing on heads would be $1/2 \times 1/2 = 1/4$. To find the probability that independent events will *both* happen, the general rule is to multiply the probabilities of the individual events.

Monohybrid cross probabilities: In the example of smooth and wrinkled seeds, heterozygotes produce *S* and *s* gametes. The probability of a gamete being *S* is 1/2. The probability of that an F_2 plant will be *SS* is $1/2 \times 1/2 = 1/4$.

Adding probabilities: The probability of an event that can occur in two or more ways is the sum of the probabilities for each way in which the event can occur. For example, the genotype *Ss* can result from *s* in the egg and *S* in the sperm, or from *S* in the egg and *s* in the sperm. Thus the probability of heterozygotes in the F_2 generation of a monohybrid cross is $1/4 + 1/4 = 1/2$.

The dihybrid cross: To calculate the probabilities of the outcomes of dihybrid crosses, simply multiply the outcomes from each of the individual monohybrid components. For example, the probability of the *SSYy* genotype can be calculated as follows: An F_1 (dihybrid) cross of *SsYy* generates 1/4 *SS*, 1/2 *Ss*, 1/4*ss*, and 1/4 *YY*, 1/2 *Yy*, 1/4 *yy*. The probability of the *SSYy* genotype is the probability of the *SS* genotype, which is 1/4, times the probability of the *Yy* genotype, which is 1/2. This would be 1/8 ($1/4 \times 1/2$).

Mendel's laws can be observed in human pedigrees. Patterns for over 2,500 inherited human characteristics have so far been determined. Humans cannot be studied using planned crosses, so human geneticists rely on pedigrees, which show phenotype segregation in several generations of related individuals. Since humans have such small numbers of offspring, human pedigrees do not show clear proportions. The actual number of affected versus unaffected offspring is impossible to predict for a certain couple because outcomes for small samples fail to follow closely the expected outcomes.

If neither parent has a given phenotype, but it shows up in the progeny, the trait is recessive and the parents are heterozygous. The chance of other children getting the trait is 1/4. Half of the children from such a cross will be carriers (heterozygous for the trait). The probability of a carrier (heterozygote) for a rare allele unknowingly marrying another unrelated carrier is quite low.

For a dominant allele, *every* affected person has an affected parent. About half of the offspring of an affected person are also affected (assuming only one parent is affected). The phenotype occurs equally in both sexes.

For a rare recessive trait, Affected people usually have parents who are both not affected. One-quarter of the children, on average, of unaffected parents would be affected. The phenotype occurs equally in both sexes.

Marriage between close relatives results in a higher likelihood that both parents will be carriers of a rare allele and produce affected children. The major use of pedigree analysis is in clinical evaluation and counseling of patients with inherited abnormalities.

3.1.4 Some Exceptions to Mendel's Laws

I. Linkage - If two different genes are located relatively close to each other on the same chromosome they can not segregate independently. The term **centi-**

morgan (cM) is used in **eukaryotic genetics** and **map unit/Morgan** in **microbial genetics**. The numbers of linkage groups are equal to number of haploid sets of chromosomes. But in XX-XY type of organisms, number of linkage groups will be one more in male than in the female as in male X and Y both are different chromosomes.

***Cis–trans* arrangement of genes:** If the dominant alleles A, B of two linked genes are present on the same chromosome and their recessive alleles are present on the homologous chromosomes, the arrangement of genes is called **cis-arrangement**. On the other hand, if one dominant gene and other recessive gene are present on the same chromosome (A,b) and their alleles type (a,B) on the homologous chromosome, this type of arrangement is called **trans-arrangement.**

$$\text{Recombination Frequency} = \frac{\text{Number of recombinants}}{\text{Total progeny}} \times 100$$

Several ascomycetes fungi (e.g., Neurospora) produce asci, which hold the haploid ascospores produced after meiosis in a specific linear order, an ordered tetrad. This order reflects the organization and chromatids involved in recombination (crossover) at meiosis. Maximum recombination frequency possible is 50%.

Interference: Crossing over takes place at chaismata. These are physical structures involving two chromatids. Not surprisingly the presence of one chaismata in a particular chromosome region can reduce the frequency of others forming close to it. This is termed as interference. The extent of interference is calculated as **coefficient of coincidence (S),** which is equal to the observed number of double crossovers divided by the expected number of double crossovers.

Coefficient of Interference is 1 minus coefficient of coincidence (1-S). If it comes to be positive values interference is termed as positive while if the value is negative it is termed as negative interference.

II. Dominance - Mendel actually investigated complete dominance, i.e. dominant gene totally masks the recessive. But in certain cases the alleles show codominant and incomplete dominant nature.

Codominance: Codominant alleles occur when rather than expressing an intermediate phenotype, the heterozygotes express both homozygous phenotypes. An example is in human ABO blood types, the heterozygote AB type manufactures antibodies to both A and B types. Blood type A people manufacture only anti-B antibodies, while blood type B people make only anti-A antibodies. Codominant alleles are both expressed. Heterozygotes for codominant alleles fully express both alleles. Blood type AB individuals produce both A and B antigens. Since neither A nor B is dominant over the other and they are both dominant over O they are said to be codominant.

Incomplete dominance: Incomplete dominance is a condition when neither allele is dominant over the other. The condition is recognized by the heterozygotes expressing an intermediate phenotype relative to the parental phenotypes. If a red flowered plant is crossed with a white flowered one, the progeny will all be pink. When pink is crossed with pink, the progeny will be 1 red, 2 pink, and 1 white.

Flower color in snapdragons is an example of this pattern. Cross a true-breeding red strain with a true-breeding white strain and the F_1 are all pink (heterozygotes). Self-fertilize the F_1 and you get an F_2 ratio of 1 red: 2 pink: 1 white. This would not happen if true blending had occurred (blending cannot explain traits such as red or white skipping a generation and pink flowers crossed with pink flowers should produce only pink flowers).

III. Multiple alleles - three or more alleles for same gene, e.g. three alleles for human blood groups A, B, O with O the recessive and A and B codominants. Human blood types exhibit both codominance and multiple alleles. B. There are about 100 alleles for hair color in mammals.

IV. Polygenic inheritance – Polygenic inheritance is a pattern responsible for many features that seem simple on the surface. Many traits such as height, shape, weight, color, and metabolic rate are governed by the cumulative effects of many genes. Such traits are called as **quantitative traits** or **polygenic traits**. Such traits are not expressed as absolute or discrete characters, as was the case with Mendel's pea plant traits. Instead, polygenic traits are recognizable by their expression as a gradation of small differences (a continuous variation). The results form a bell shaped curve, with a mean value and extremes in either direction.

Height in humans is a polygenic trait, as is color in wheat kernels. Height in humans is not discontinuous. If you line up the entire class a continuum of variation is evident, with an average height and extremes in variation very short [vertically challenged] and very tall [vertically enhanced]. Traits showing continuous variation are usually controlled by the additive effects of two or more separate gene pairs. This is an example of polygenic inheritance. The inheritance of each gene follows Mendelian laws.

Usually polygenic traits are distinguished by

1. Traits are usually quantified by measurement rather than counting.
2. Two or more gene pairs contribute to the phenotype.
3. Phenotypic expression of polygenic traits varies over a wide range.

Human polygenic traits include: Height, Weight, Eye color, Intelligence, Skin color (pigmentation, Figure 3.6), and many forms of behavior. Individual heritable characters are often found to be controlled by groups of several genes, called polygenes. Each allele intensifies or diminishes the phenotype. Variation is continuous or quantitative ("adding up"). Examples of continuous characters are height, skin color, and possibly intelligence.

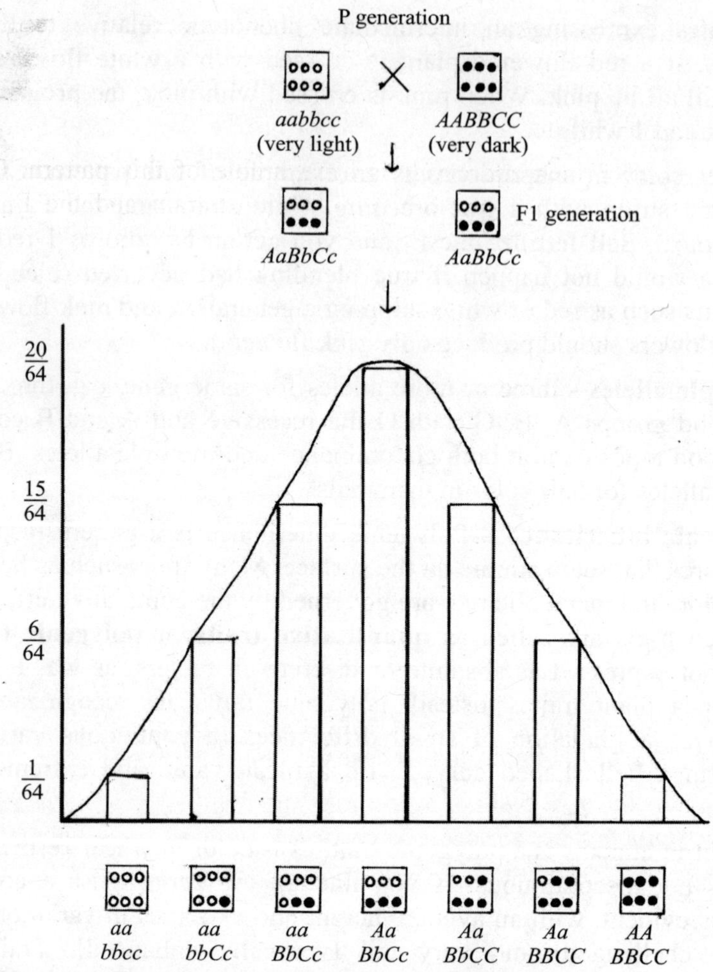

Figure 3.6. Polygenic inheritance. Showing skin pigmentation in humans

V. Pleiotropy – Some alleles have multiple phenotypic effects

Pleiotropy is the effect of a single gene on more than one characteristic (phenotype). An example is the "frizzle-trait" in chickens. The primary result of this gene is the production of defective feathers. Secondary results are both good and bad; good include increased adaptation to warm temperatures, bad include increased metabolic rate, decreased egg-laying, changes in heart, kidney and spleen. Cats that are white with blue eyes are often deaf, white cats with a blue and an yellow-orange eye are deaf on the side with the blue eye. Sickle-cell anemia is a human disease originating in warm lowland tropical areas where malaria is common. Sickle-celled individuals suffer from a number of problems, all of which are pleiotropic effects of the sickle-cell allele. Another example is the coloration pattern and crossed eyes of Siamese cats, which are both caused by the same allele. These unrelated characters are caused by the same protein produced by the same allele.

VI. Environment may affect the expression of genes. Examples

Eg1. Siamese cats have dark ears, nose and paws because they are colder than rest of the body. The enzyme responsible for the dark color only functions at lower temperatures (recall effect of temperature on enzymes).

Eg2. Above water versus submerged leaves on some plants are markedly different.

Eg3. Identical twins have identical genes but exercise, diet and a number of other factors may cause them to have different phenotypes.

VII. Cytoplasmic inheritance - recall that plastids and mitochondria have their own DNA. Therefore some characteristics may be under the control of genes in the cytoplasm.

Eg1. In plants cytoplasmic inheritance from **plastids** produces the variegated leaves of some leaves, e.g. Coleus.

Eg2. Male sterility in some cultivated plants is due to a **mitochondrial gene**. This is useful for producing artificial hybrids because the stamens do not have to be removed to prevent self-pollination.

VIII. Sex chromosomes - in many organisms, particularly animals, the males and females may have one pair of chromosomes that are different and they determine the sex of the individual. These are called sex chromosomes.

A. Humans have 22 pairs of autosomes and one pair of sex chromosomes. In humans the male, Y chromosome, carries a master gene called SRY (Sex determining Region of the Y chromosome) which causes the development of male hormones and reproductive organs.

3.2 POPULATION GENETICS

The origin of species convinced most biologists that species are products of evolution, but Darwin was not nearly so successful in gaining acceptance for his idea that natural selection is the mechanism of evolution. Natural selection requires hereditary processes that Darwin could not explain. The main thing missing in Darwin's theory was an understanding of inheritance that could explain how chance variations arise in a population and their precise mechanism of transmission to successive generation. Although Gregor Mendel and Charles Darwin were contemporaries, Mendel's discoveries were unappreciated at the time, and apparently no one noticed that he had elucidated the very principles of inheritance that could have resolved Darwin's paradox and given credibility to natural selection.

3.2.1 Integration of Darwinian selection and Mendelian inheritance to give rise to modern evolutionary theory

Ironically, when Mendel's research articles was rediscovered and reassessed at the beginning of the twentieth century, many geneticists believed that the laws of inheritance were at odds with Darwin's theory of natural selection, Darwin emphasized quantitative characters, those characteristics in a population that vary along a continuum, such as human height, human skin color or the speed with which an animal can flee from a predator. Today it is well known that quantitative traits are governed by multiple genetic loci. But Mendel recognized only discrete "either-or" traits, such as purple or white flowers in pea plants, as heritable. Thus, there seemed to be no genetic basis for natural selection to work on the more subtle variations within a population that were central to Darwin's theory.

An important turning point for the evolutionary theory was the birth of **population genetics,** which emphasizes the extensive genetic variations within populations and recognizes the importance of quantitative characters. With progress in population genetics in the 1930s, Mendelism and Darwinism were reconciled, and the genetic basis of variation and natural selection was worked out. With the advent of **modern synthesis** in 1940s, emphasis was given to population as a unit of evolution and natural selection as the important mechanism of evolution.

3.2.2 Genetic structure of a population

Population genetics is the study of alleles of genes in populations, and the forces, which maintain or change the frequencies of particular alleles and genotypes in populations. A **population** is a localized group of individuals belonging to the same species. For now, we will define a **species** as a group of populations whose individuals have the potential to interbreed and produce fertile offsprings in nature. Each species has a geographical range within which individuals are unevenly distributed, but are usually concentrated in several localized populations. A population may be isolated from others of the same species, exchanging genetic material only rarely. Such isolation is particularly common for populations confined to widely separated islands, unconnected lakes or mountain ranges separated by low lands; such local interbreeding population is often called as **deme**. A deme may be difficult to define, for example, humans living in one area may actually marry partners chosen on criteria of religion, race, wealth or social class. Similarly, insects on trees in a forest may mate with others on the same tree, and rarely travel to the next tree to mate. The insect population may fill the forest, or each tree may have its own semi-isolated population. Natural populations are composed of many interbreeding individuals, each with a unique combination of genes and alleles, but the population shares a **gene pool**. Evolution is the change of frequencies of alleles in the gene pool, so population genetics is of fundamental importance. Some genotypes, such as those associated with human genetic diseases, are rare, and population genetics is important for understanding why this is so, and for predicting changes in the incidence of such diseases. Most individuals have phenotypes close to the average, but some carry rare combinations of genes or alleles which cause their phenotype to deviate markedly from a typical member of the species.

Gene pool is the total aggregate of genes in a population at any one time. It consists of all alleles at all genetic loci in all individuals of the population. In a diploid species each locus is represented twice in the genome of an individual either as homozygous or heterozygous for those homozygous loci. Each homozygous individual have two identical alleles for a given gene whereas heterozygous individuals have two different alleles for that gene or character. If all the members of a population are homozygous for the same allele, that allele is said to be fixed in the gene pool. Usually, there are two or more alleles for a gene, each having a relative frequency in the gene pool.

The main goal of population genetics is to study the **genetic structure** of a local population. The genetic structure of a population includes the allele frequency and the genotypic frequency of a population gene pool. An example will make the concept of allele frequency and genotypic frequency more clear. Imagine a wild flower plant population with two varieties contrasting in flower color. An allele for purple flowers, which we will symbolize by *A*, is completely dominant to an allele for white flowers, symbolized *a*. For sake of simplicity, these are the only two alleles for this locus (flower color) in the population. Suppose an imaginary population has 1000 plants, and 40 of these plants have white flowers because they are recessive for homozygous alleles; their genotype is *aa*. The other 960 plants have purple flowers; 640 are homozygous *(AA)* and 320 are heterozygous *(Aa)*. Because these are diploid organisms, there are a total of 2000 copies of genes for flower color in the population of 1000 individuals. The dominant allele *(A)* accounts for 1600 of these genes (640 × 2 = 1280 for *AA* plants, plus 320 X 1 = 320 for *Aa* individuals). Thus the frequency of the *A* allele in the gene pool of this population is 1600/ 2000 = 0.8 = 80%. And there are only two allelic forms of the gene, the *a* allele must have a frequency of 0.2, or 20%.In the same way the genotypic frequency is the total number of individuals of a particular genotype from the total numbers of all the individuals in a population. In our imaginary wildflower plant population. The genotypic frequency of *AA* = 0.64 (640/1000), *Aa* = 0.32 (320/1000) and *aa* = 0.04 (40/1000).

3.2.3 Hardy-Weinberg theorem for non-evolving populations

Before we consider the mechanisms that cause a population to evolve, it will be helpful to examine, for comparison, the genetic structure of a non-evolving population. Such a gene pool is best described by the **Hardy-Weinberg theorem**, named after the two scientists who derived the principle independently in 1908. *It states that the frequencies of alleles and genotypes in a population's gene pool remains constant over the generation unless acted upon by agents other than sexual recombination.* Put another way, the sexual shuffling of the alleles due to meiosis and random fertilization has no effect on the overall genetic structure of a population. To apply the Hardy-Weinberg theorem, let's return to our imaginary wild flower population of 1000 plants (Figure 3.7a). In our wild flower plant population, 80% (0.8) of the flower-color loci in the gene pool have the *A* allele and 20% (0.2) have the *a* allele. How will genetic recombination during sexual reproduction affect the frequencies of the two alleles in the next generation of our wildflower population? We will assume that the union of sperm and ova in the population is completely random; that is, all male-female mating combination are equally likely. The situation is just analogous to mixing all gametes in a box and then drawing them randomly, two at a time, to determine the genotype for each zygote (fertilized egg). Each gamete

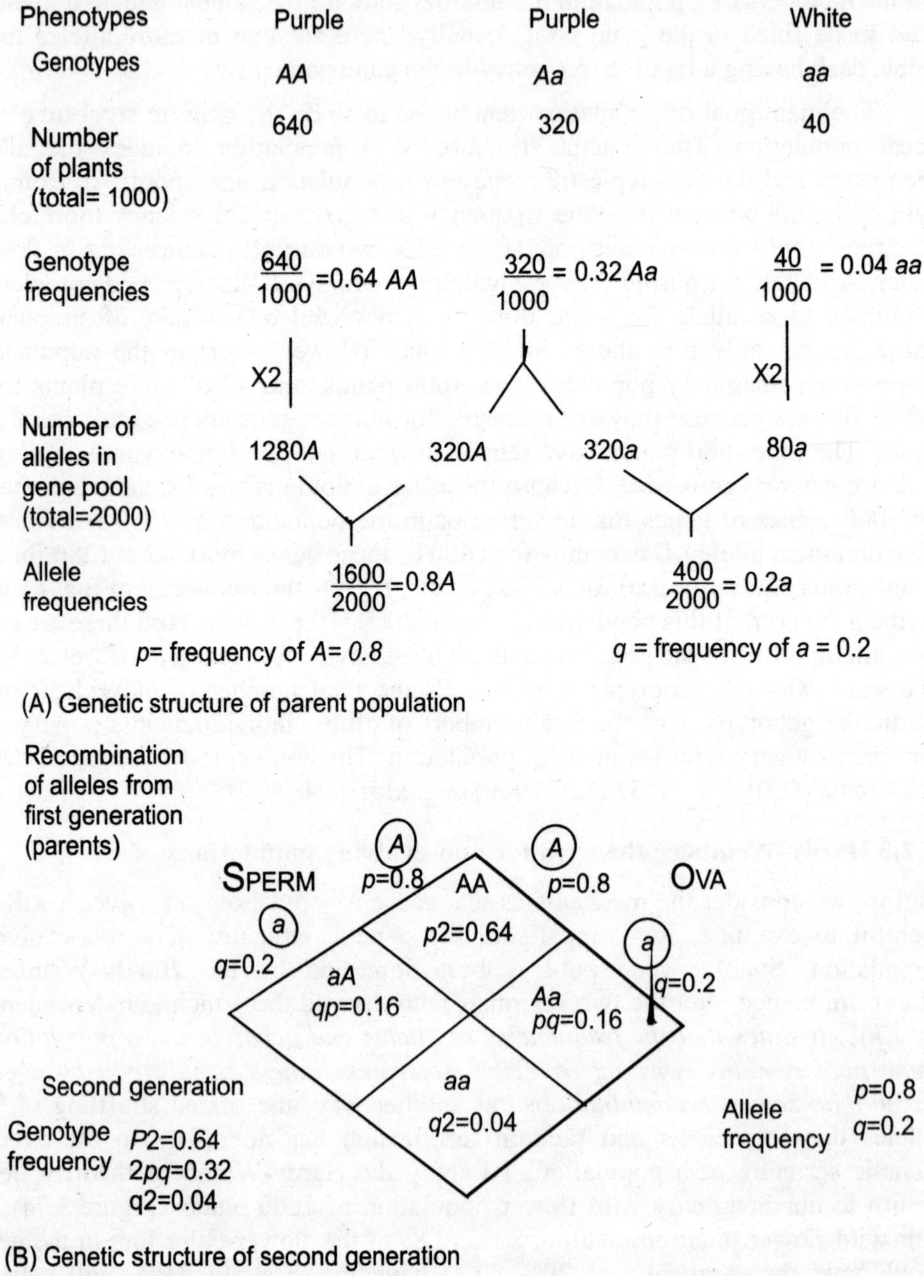

Figure 3.7. The Hardy-Weinberg theorem

has one allele for flower color, and the allele frequencies of the gametes will be the same as the allele frequencies in the parent population. Every time a gamete is drawn from the pool at random, the chance of that gamete will bear an *A* allele is 0.8, and the chance that the gamete will have an *a* allele is 0.2.Using the rule of multiplication, we can calculate the frequencies of the three possible genotypes in the next generation of the population (Figure 3.7b). The probability of picking two *A* alleles from the pool of gametes is 0.64 (0.8×0.8). Thus, about 64% of the plants in the next generation will have the genotype *AA*. The frequency of *aa* individuals will be about 4%, or 0.04 (0.2×0.2). And 32%, or 0.32, of the plants will be heterozygous, i.e. *Aa* or *aA,* depending on whether sperm or ovum that supplies the dominant allele. It is clear from Figure 3.7 that the sexual processes of meiosis and random fertilization have maintained the same allele and genotype frequencies that existed in the previous generation of the wildflower population. For the flower color locus, the population's genetic structure is in a state of equilibrium referred to as **Hardy-Weinberg equilibrium.**

3.2.4 Hardy-Weinberg equilibrium for two loci

We can use our imaginary wildflower population to describe the Hardy-Weinberg theorem in more general terms. For a gene locus where only two alleles occur in a population, population geneticists use the letter p to represent the frequency of one allele and the letter q to represent the frequency of the other allele. In our imaginary wildflower population, $p = 0.8$ and $q = 0.2$. Note that $\boldsymbol{p + q = 1}$; the combined frequencies of all possible alleles must account for 100% of the genes for that locus in the population. This equation is Hardy-Weinberg equation for allele frequency at two loci. If there are only two alleles and we know the frequency of one allele, the frequency of the other can be calculated:

If $\quad p + q = 1 \quad$ then $\quad p = 1 - q \quad$ and $\quad q = 1 - p$.

When gametes combine their alleles to form zygotes, the probability of generating an *AA* genotype is p^2 (an application of the rule of multiplication). In our wildflower population, $p = 0.8$, and $p^2 = 0.64$, the probability of an *A* sperm fertilizing an *A* ovum to produce an *AA* zygote. The frequency of individuals homozygous for the other allele (*aa*) is q^2, or $0.2 \times 0.2 = 0.04$ for the wildflower population. There are two ways in which an *Aa* genotype can arise, depending on which parent contributes the dominant allele. Therefore, the frequency of heterozygous individuals in the population is $2pq$ ($2 \times 0.8 \times 0.2 = 0.32$). if we have included all possible genotypes for two loci, the genotype frequencies add up to 1.

The Hardy-Weinberg equation for genotypic frequency at two loci can be expressed as:

$$p^2 \quad + \quad 2pq \quad + \quad q^2 \quad = \quad 1$$

For our wildflowers, this is $0.64 + 0.32 + 0.04 = 1$.

The Hardy-Weinberg equation enables us to calculate frequencies of alleles in a gene pool if we know frequencies of genotypes, and vice versa. If the genetic structure of a population is constant for successive generation then that population is non-evolving, but if the genetic structure of population changes from one generation to next than the population is said to be evolving, so with the help of Hardy-Weinberg equation we can describe both non-evolving and evolving populations. Another important application of Hardy-Weinberg equation is to calculate the percentage of the human population that carries the alleles for a particular inherited disease. For example, consider an inherited disorder, say Sickle cell anemia, a genetic disorder which may cause death due to abnormal (Sickle shaped) RBCs. The disease is caused by a recessive allele. Consider a human population of any state/country in which there are 10,000 individuals and it was found that out of 10,000 individuals, 400 individuals are diseased (Sickle cell anemia). Now from this data we can calculate (a) allele frequency of the diseased/mutant/recessive allele (q), (b) allele frequency of normal/dominant allele (p), (c) genotypic frequency of normal individuals (p^2), (d) genotypic frequency of carriers ($2pq$) and (e) genotypic frequency of diseased individuals (q^2).

Since total number of diseased individuals is given as 400

Therefore $q^2 = 400/\ 10{,}000 = 0.04$ or $q = \sqrt{0.04} = 0.2$ or 20%

Now $q = 0.2$ then $p = 1 - q$ or $p = 1 - 0.2 = 0.8$ or 80%

$p^2 = p \times p$ or $0.8 \times 0.8 = 0.64$ or 64%

$q^2 = q \times q$ or $0.2 \times 0.2 = 0.04$ or 4%

$2pq = 2 \times p \times q$ or $2 \times 0.8 \times 0.2 = 0.32$ or 32%.

From the above calculation it is clear that 20% of the above mentioned human population carries the diseased allele and 80% of the population is carrying the normal allele. Further it is clear that 64% of the population i.e. 6400 individuals are normal 32% of the population i.e. 3200 individuals are carriers and 4% of the population i.e. 400 individuals are diseased.

3.2.5 Hardy-Weinberg equilibrium for multi-loci

Hardy-Weinberg equilibrium can be implemented even for genes or genetic loci having more than two alleles (three, four or more). In this reference we are taking three alleles A_1, A_2 and A_3, and define their frequencies as p, q, and r, respectively. We form new zygotes by sampling two successive gametes from a large pool of gametes. The first gamete we pick could be A_1, A_2 or A_3. If we first pick (with chance p) an A_1 allele from the gamete pool, the chance that the second allele is another A_1 allele is p, the chance that it is an A_2 allele is q, and the chance that it is an A_3 allele is r: from these three, the frequencies of A_1A_1,A_1A_2, and A_1A_3 zygotes are p^2 , pq, and pr.

Now suppose that the first allele we picked out had been an A_2 (which would happen with chance q). The chances that the second allele would again be A_1, A_2 or A_3 would be p, q, and r, respectively, giving A_1A_2,A_2A_2, and A_2A_3 zygotes in frequency pq, q^2, and qr. Finally, if we had picked (with chance r) an A_3 allele, we produce A_1A_3,A_2A_3, and A_3A_3 zygotes in frequency pr, qr, and r^2.

The only way to find homozygotes A_1A_1,A_2A_2, and A_3A_3 is by picking two of the same kind of gametes whose frequencies are p^2, q^2, and r^2. The heterozygotes can be formed from more than one kind of first gamete and their frequencies are obtained by addition. The total chance of forming an A_1A_3 zygote is $pr + rp = 2pr$; of forming an A_1A_2 zygote is $pq + qp = 2pq$; and of an A_2A_3 zygote is $2qr$.

The complete Hardy-Weinberg equation for genotypic frequency for three loci is:

$$P^2 + 2pq + 2pr + q^2 + 2qr + r^2 = 1$$

The complete Hardy-Weinberg equation for allele frequency for three loci is:

$$P + q + r = 1$$

3.2.6 Conditions required to maintain Hardy-Weinberg equilibrium

There are five conditions that are required for Hardy-Weinberg equilibrium to be maintained in a non-evolving population:

1. *Very large population size.* In a small population, genetic drift, which is chance fluctuation in the gene pool, can alter the frequencies of alleles and thus the genetic structure of the population. So if such type of fluctuations occurs in large size populations then there will be no effect on the overall genetic structure of the population and it will remain in equilibrium.
2. *Isolation from other population*. Gene flow, the transfer of alleles between populations due to the movement of individuals or gametes, can change the gene pool. So it is a must that the population should be isolated with other population of the same species to prevent any gene flow, or in other words population should be a closed system in terms of gene flow to maintain Hardy-Weinberg equilibrium.
3. *No net mutations*. We know that mutations are the ultimate source of discontinuous variations, they can alter the gene pool and even they can generate a new allele in the gene pool. So for a population in Hardy-Weinberg equilibrium it should be far away from mutations.
4. *Random mating*. If the individuals select mates having certain inheritable traits, then the random mixing of gametes required for Hardy-Weinberg equilibrium does not occur. So in a natural population every individual has equal opportunity to mate with another to maintain the Hardy-Weinberg equilibrium.
5. *No natural selection*. According to Darwin, natural selection is the main driving force of evolutionary changes and it operates by means of differential survival and differential reproduction. Whenever natural selection come into play it will alter a gene pool by favouring the transmission of some alleles at the expense of others.

So the above five conditions must be strongly followed for calling a population as non-evolving and if any one of the above mentioned factor does not came into play then the population is no longer in Hardy-Weinberg equilibrium and it will be considered as an evolving population. Since in real condition these five factors are never applicable strictly on any type of population of whichever species it may be, therefore, it can be considered that in real condition there is hardly any population in the living world which is non-evolving or in other words all the Mendelian populations are evolving.

3.3 MICROEVOLUTION AND ITS CAUSES

If the Hardy-Weinberg theorem describes a gene pool in equilibrium, that is, a nonevolving population, then how is it relevant to the study of evolution. The concept of Hardy-Weinberg equilibrium tells us what to expect if a population is not evolving. The equilibrium values for allele and genotype frequencies we calculate from the Hardy-Weinberg equation provide a baseline for tracking the genetic structure of a population over a succession of generations. If the frequencies of alleles or genotypes deviate from values expected from Hardy-Weinberg equilibrium, then the population is evolving. We can now define evolution at population level as: *Evolution is a generation-to-generation change in population's frequencies of alleles or genotypes, a change in a populations genetic structure.* Because such change in a gene pool is evolution on the smallest scale, it is referred to more specifically as **microevolution.**

Microevolution is the evolution at the smallest level and it operates below species level. Microevolution is occurring even if the frequencies of alleles are changing for only a single genetic locus. It is possible that if we track allele and genotypic frequencies in a population over successive generations, some loci may be at equilibrium while allele frequencies at other loci are changing. Such a population is evolving. For example, our imaginary wildflower population would be evolving if the frequencies of the purple flower and white flower alleles were changing from generation to generation, even if Hardy-Weinberg equilibrium were maintained for all other loci.

Causes of micro evolutionary changes

There are five main causal agents for microevolution:

- Genetic drift
- Gene flow
- Mutations
- Nonrandom mating
- Natural selection.

In this segment we will describe each causal agent and how it is responsible for micro-evolutionary changes.

3.3.1 Genetic drift

If we flip a coin thousand times, and a result of 700 heads and 300 tails would make you very suspicious about that coin. Flip a coin ten times, and an outcome of seven heads and three tails is within reason. The smaller a sample, the greater the chance of deviations from the expected result – an equal number of heads and tails, in the case of coin tosses. This disproportion of results in a small sample is known as sampling error, and it is an important factor in the genetics of small populations of organisms. If a new generation draws its allele at random, then the larger the sample size, the better it will represent the gene pool of the previous generation. If a population of organism is small, its existing gene pool may not be accurately represented in the next generation because of sampling error and this sampling error is because the genes that form a new generation are a *random sample* from the parental generation. Random sampling starts at conception. In every species, each individual produces many more gametes than will ever fertilize, or be fertilized, to form new organisms. The successful gametes which do form offspring are a sample from the many gametes that the parents produce. If a parent is homozygous, the sampling makes no difference to what genes end up in the offspring; all of a 'homozygote's gametes contain the same gene. However, sampling does matter if the parent is heterozygote, such as *Aa*. It will then produce a large number of gametes, of which approximately one-half will be *A* and the other half *a* (The proportion may not be exactly one-half. Reproductive cells may die at any stage leading to gamete formation, or after they have become gametes; also, in the female, a randomly picked three-quarters of the products of meiosis are lost as polar bodies). If that parent produces 10 offspring, it is most likely that five will inherit an *A* gene and five an *a* gene. But because the gametes that formed the offspring were sampled from a much larger pool of gametes, it is possible that the proportions would be something else. Perhaps six inherited *A* and only four *a*, or three inherited *A* and seven *a*.

In the small population of wildflowers, for example, the frequencies of the alleles for purple (*A*) and white (*a*) flowers fluctuate over several generations, genetic variations has been reduced. This is a case of microevolution caused by random changes in the gene pool of small population due to chance and it is called as **genetic drift** or **random drift**.

Genetic drift may be adaptive in nature if it favours the more number of normal alleles or genes in the next generation; it may be non-adaptive if it favours the mutant alleles or genes in the next generation of a small population.

Ideally, a population must be infinitely large for genetic drift to be ruled out completely as an agent of evolution. Although that is impossible, many populations are so large that drift may be negligible. However, some populations are small enough for significant genetic drift to occur. Two situations that can lead to genetic drift are the population bottlenecks and the founding of new colonies by a small number of individuals (Founder effect).

The Bottleneck effect

Drastic short-term unselective reductions of population size caused by natural disasters (earthquakes, flood, or fire), disease, or predators may result in (by chance) the survivors representing only a small portion of the original gene pool. By chance, certain alleles will be overrepresented, and some alleles may be eliminated completely (Figure 3.8). Genetic drift may continue to affect the population for many generations, until the population is again large enough for sampling errors to be insignificant. Even when the population increases to its original size, a portion of its original genetic diversity remains lost. This feature, termed bottleneck, is a problem with many endangered species. A bottleneck effect is genetic drift in which a severe reduction in population size results from natural disaster, predation, or habitat reduction. This results in a severe reduction of the total genetic diversity of the original gene pool because alleles for at least some loci are likely to be lost from the gene pool. One example concerns the population of northern elephant seals, which passed through a bottleneck in the 1890s when hunters reduced it to about 20 individuals. Since then, the animal has become a protected species, and the population has grown to over 30,000. Researchers have examined 24 gene loci in many individuals of the northern elephant seal population, and no genetic variation has been found; a single allele has been fixed at each of the 24 loci, probably due to genetic drift. The cheetah bottleneck causes relative infertility because of the intense inbreeding. Similarly, the Hawaiian silversword has passed recently through its pawn bottleneck. Recent studies on humans suggest that there may have been one or more instances of severe genetic bottlenecks in our own prehistory. The bottleneck effect prevents most genotypes from participating in production of next generation.

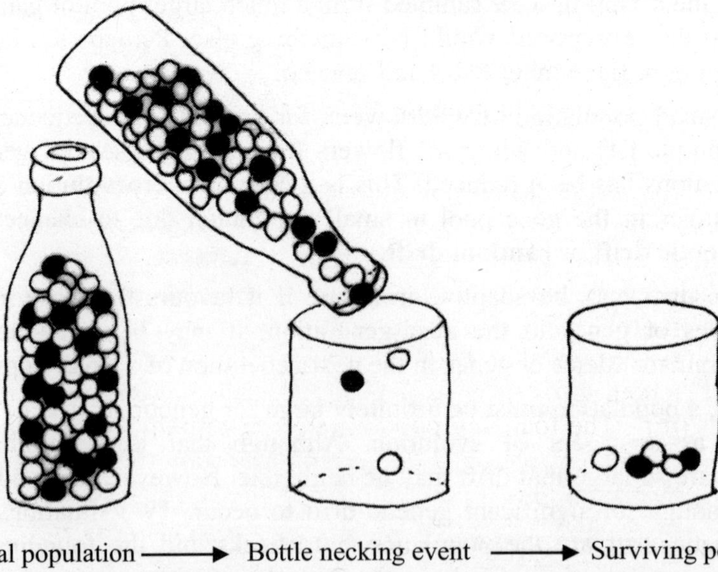

Original population ——→ Bottle necking event ——→ Surviving population

Figure 3.8. The bottleneck effect. A gene pool can begin to drift by chance after the population is drastically reduced by a disaster that kills unselectively. Notice that the survivors will represent the gene pool of that population.

The Founder effect

Genetic drift is also likely whenever a few individuals colonize an isolated island, lake, or some other new habitat. The smaller the sample size, the less the genetic makeup of the colonists will represent the gene pool of the larger population they left. The most extreme case would be the founding of a new population by one pregnant animal or a single plant seed. If the colony is successful, random drift will continue to affect the frequency of alleles in the gene pool until the population is large enough for sampling errors from generation to generation to be minimal. Genetic drift in a new colony is known as **Founder effect.** The effect undoubtedly contributed to the evolutionary divergence of Darwin's finches after strays from the South American mainland reached the remote Galapagos Islands. We can divide the definition into two parts. The first part is establishment of a new population by a small numbers of founders; we can call that a "Founder event." The second part is that the founders have a limited sample of genetic variation. The full Founder effect requires not only a founder event, but also that the founders are genetically unrepresentative of the original population.

Founder events undoubtedly happen. A population may be descended from a small number of ancestral individuals for either two reasons. A small number of individuals may colonize a place previously uninhabited by their species; the 250 or so individuals making up the modern human population on the island of Tristan da Cunha, for example, are all descended from about 20–25 immigrants in the early nineteenth century, and most are descended from the original settlers - one Scotchman and his family, who arrived in 1817. Alternatively, a population that is established in an area may fluctuate in size; the founder effect then occurs when the population passes through a bottleneck in which only a few individuals survive, and later expands again when more favourable times returns.

Founder events are unlikely to produce homozygosity

An interesting thing is that if we take a small sample of individuals from a larger population then what is the chance that they will have reduced genetic variations or in other words what is the chance that an allele is lost from the small sample. In the special case of two alleles (A and a with proportions p and q), if one of them is not included in the founder population, the new population will be genetically monomorphic. The chance that an individual will be homozygous AA is simply p^2. The chance that two individuals drawn at random from the population will be both AA is $(p^2)^2$; in general, the chance of drawing N identical homozygotes is $(p^2)^N$. The founding population could be homozygous either it is made up of N AA homozygote or N aa homozygotes, and the total chance of homozygosity is therefore:

$$\textit{Chance of homozygosity} = [(p^2)^N + (q^2)^N]$$

3.3.2 Gene flow

Hardy-Weinberg equilibrium requires the gene pool to be a closed system, but most populations are not completely isolated. A population may gain or lose alleles by gene flow, genetic exchange due to migration of fertile individuals or gametes between populations. Perhaps, for example, a population near our hypothetical wild flower population consists entirely of purple-flowered individuals (*AA*). A wind storm may blow pollen from *AA* population to our wildflower population, and allele frequency may change in next generation. Gene flow tends to reduce differences between populations that have accumulated because of natural selection or genetic drift. If it is expensive enough, gene flow can eventually amalgamate neighboring populations into a single population with a common genetic structure. As humans began to move about the world freely, gene flow undoubtedly became an important agent of microevolutionary change in populations that were previously quite isolated.

3.3.3 Mutations

A mutation is the structural, physical or any chemical change in an organism's DNA. If a new mutation occurs in a gamete and if this gamete fuses to form a new organism than it can immediately change the gene pool of a population by substituting one allele for another. For example, a mutation that causes a purple-flowered plant (*AA*) in our hypothetical wildflower population to produce gametes bearing the mutant allele for white flowers (*a*) would decrease the frequency of the allele *A* allele in the population and increase the frequency of the *a* allele.

For any one gene locus, however, mutation alone does not have much quantitative effect on a large population in a single generation. This is because a mutation at any given gene locus is a very rare event; although mutation rates vary, depending on the species and the gene locus, rates of one mutation per locus per 10^5 to 10^6 gametes are typical. If an allele has a frequency of 0.50 in the gene pool and mutates to another allele at a rate of 0.00001 mutations per generation, it would take 2000 generations to reduce the frequency of the original allele from 0.50 to 0.49. If some new allele produced by mutation increases its frequency by a significant amount in population, it is not because mutation is generating the allele in abundance, but because individuals carrying the mutant allele are producing a disproportionate number of offspring as a result of natural selection or genetic drift.

Although mutations at a particular gene locus are rare, the cumulative impact of mutations at all loci can be significant. This is because each individual has thousands of genes, and many populations have thousands or millions of individuals. Certainly over the long term, mutation is, in itself, very important to evolution because it is the original source of the genetic variation that serves as the raw material for natural selection.

3.3.4 Non-random mating

The Hardy-Weinberg equilibrium requires random mixing of gametes and it is possible when an individual of any genotype select its mate at random from the population. But in reality individuals rarely mates randomly, rather individuals prefer to mate more often with close neighbors than with more distant members of the population, especially in species that do not disperse far. Other individuals in the same "neighborhood" within a larger population tend to be closely related. This promotes **inbreeding,** mating between closely related partners. The most extreme case of inbreeding is selfing (self-fertilization), particularly common in plants. Although the effect of inbreeding will not change the overall gene frequency, it will lead to an excess of homozygous genotypes. Inbreeding will thus cause a rare recessive allele to appear in greater homozygous frequency than under random mating, offering increased opportunity for selection to act on rare recessives. For example, in our imaginary wildflower population, self-pollination would tend to increase the frequency of homozygous genotypes at the expense of heterozygotes. If *AA* individuals and *aa* individuals "self," then their offspring must also be homozygous. If *Aa* plants "self," however, only half of their offspring will be heterozygous. With each generation, the proportion of heterozygotes decreases and the proportion of dominant and recessive homozygotes increases (Figure 3.9).

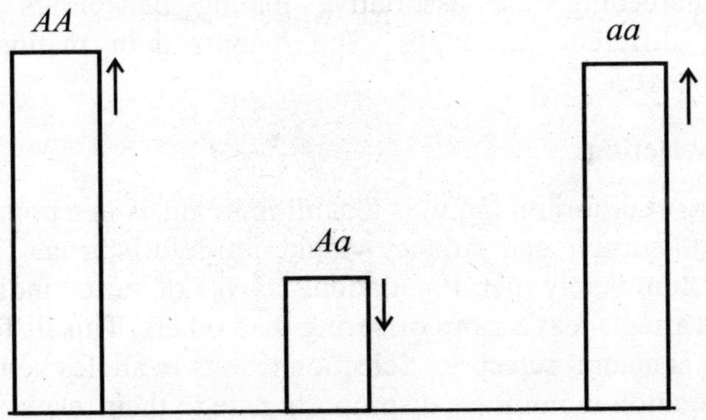

Figure 3.9. The consequences of inbreeding – there is increase in homozygosity (*AA* and *aa*) and decrease in heterozygosity (*Aa*).

Even in less extreme cases of inbreeding without selfing, the proportion of heterozygotes decreases, though more slowly. One visible effect of this change in genotype frequencies is a greater proportion of individuals expressing recessive phenotypes; the frequency of white-flowered individuals in our wildflower population would be greater than the Hardy-Weinberg equation predicts. Regardless of the impact of inbreeding on the ratio of genotypes and phenotypes in the population, the frequencies of the two alleles remain the same. It is just that a smaller proportion of recessive alleles are masked in heterozygous individuals. Although some amount of inbreeding occurs in most outbreeding populations, significant amount of inbreeding can cause **inbreeding depression** in which rare deleterious recessives may now appear with increased homozygous frequency.

Another type of nonrandom mating is **phenotypic assortative mating**, in which individuals select partners that are like themselves in some phenotypic characters. In many human societies, mates are chosen that share characteristics such as height, color, facial form, muscular build, and intelligence. In such phenotypic assortative matings homozygosity can also increase, but only for those loci involved in the traits on which the preferred matings are based. In case of **genetic assortative mating**, mating is based on the genetic relationship. For example, brother-sister mating and first-cousin mating. Such type of mating system is common in animals. In contrast to phenotypic assortative mating, in genetic assortative mating homozygosity is increased at all loci. A further type of mating practice is **disassortative mating** in which individuals of unlike genotype and phenotype form mating pairs, thereby preventing inbreeding and helping to maintain heterozygosity.

Notice again that nonrandom mating - inbreeding or assortative mating - increases the number of gene loci in the population that are homozygous, but nonrandom mating does not in itself alter the overall frequencies of alleles in a population's gene pool. Remember, however, that a population's genetic structure is defined by its frequencies of alleles and genotypes. Any change in a population's inbreeding and assortative mating behaviours will shift the frequencies of different genotypes. Thus, nonrandom mating can cause a population to evolve.

3.3.5 Natural selection

Hardy-Weinberg equilibrium requires that all individuals in a population be equal in their ability to survive and produce viable, fertile offspring. This condition is probably never completely met. Populations consist of varied individuals, and on average, some variants leave more offspring than others. This differential success in reproduction is natural selection. Selection results in alleles being passed along to the next generation in numbers disproportionate to their relative frequencies in the present generation. For example, in our imaginary wildflower population, plants with purple flowers (*AA* or *Aa* genotypes) may for some reason produce more offspring on average than plants having white flowers (*aa*); perhaps white flowers are more visible to herbivorous insects that eat the flowers. This would disturb Hardy-Weinberg equilibrium; the frequency of the *A* allele would increase, and the frequency of the *a* allele would decline in the gene pool.

Of all the agents of microevolution discussed so far, only natural selection is likely the only agent who is adaptive in value, that is, it adapts a population to its environment. Natural selection accumulates and maintains favorable genotypes in a population. If the environment should change, selection responds by favoring genotypes adapted to the new conditions. But the degree of adaptation can be extended only within the realm of the genetic variability present in the population.

3.4 MACROEVOLUTION

Macroevolution is evolution at a larger scale. It involves the kinds of events that can be studied in the fossil records, such as the origin of new organs, or body plans, or of new higher taxa above species level, such as the evolution of mammal-like reptiles into mammals, fish into tetrapods, and green algae into vascular plants. It also includes diversification, extinction and replacement of higher taxa. These large-scale changes can be distinguished from microevolution, which refers to changes in gene frequencies within a population, under the influence of natural selection and random drift. The conventional dividing line between macro- and microevolution is at speciation, so that events below that level are microevolution and those above it are macroevolution.

The distinction between micro- and macroevolution has traditionally been not only between the timescales of the events but also between the methods used, microevolution has been studied with genetic techniques, and has used observation and experiment on the timescale of human lifetimes. Macroevolution has been studied with fossil evidence, comparative morphology, and phylogenetic inference.

As we know how microevolution proceeds, so it is noteworthy to explain how macroevolution proceeds. Many examples suggest that macroevolution is actually a microevolutionary process that persist for a long time, or in other words macroevolution is due to "extrapolated" microevolution (Figure 3.10). The origin of mammals is a classic example of macroevolution.

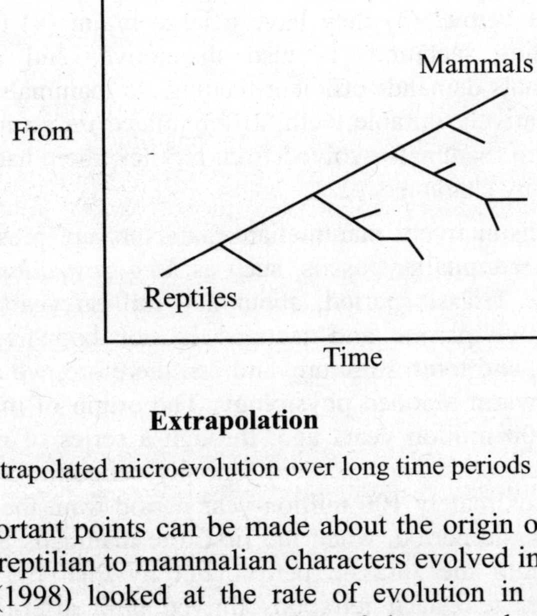

Figure 3.10. Extrapolated microevolution over long time periods

Two important points can be made about the origin of mammals. First, the changes from reptilian to mammalian characters evolved in gradual stages. Sidor and Hopson (1998) looked at the rate of evolution in mammal-like reptiles quantitatively. They measured the number of character changes per unit time and found that not only mammals evolved in many stages but also that the rate of morphological evolution was approximately constant over a 100 million-year period. The second is that the large-scale differences between mammals and reptiles concern adaptations. Mammals have a high-energy, high metabolic rate kind of physiology, with locomotory adaptations for rapid movements and adaptations for powerful and efficient feeding. These are surely adaptive changes, which would have been brought about by natural selection.

The general evolutionary model suggested by the mammal-like reptiles, therefore is one of cumulative action of natural selection over a long (100 million year) period. The accumulation of many small-scale changes resulted in the large-scale change from reptile to mammal. The theory of mammalian origins is therefore an extrapolative theory. A similar conclusion could be drawn about the origin of humans, and terrestrial plant and vertebrates. In these examples, macroevolution proceeds by the same process-natural selection and adaptive improvement- as has been observed within species and speciation; but the process is operating over a much longer period. The extrapolative model is not the only model for evolution of some major groups, but it is the most important one and the only one that can be illustrated with detailed fossil evidence. Let's take an example of macroevolution, how mammals evolves from reptiles in the next section.

3.4.1 Mammal's evolution from reptilian ancestors: a closer look

Mammals are a distinct group of vertebrates in many respects: (1) they are warm blooded and have a constant body temperature with high metabolic rate; (2) they have a characteristic mode of locomotion or gait, in which the body is held upright with the legs below; (3) they have a large brain; (4) their method of reproduction, including lactation, is also distinctive; and (5) the active metabolism of mammals demands efficient feeding, so mammals have powerful jaws and a set of relatively durable teeth, differentiated into a number of tooth types. Therefore, when mammals evolved from reptiles, there had to be changes on a large scale in many characters.

Not all of the distinctively mammalian characters are preserved in fossil records. The earliest mammalian fossils, such as *Megazostrodon* (Figure 3.11), date back to the late Triassic period, about 200 million years ago. Whether *Megazostrodon* was viviparous and lactated is not known. But it had a mammalian jaw, gait, and tooth structure, and can therefore, we can infer that it probably also had a warm-blooded physiology. The origin of mammals can be traced back before 200 million years ago, through a series of reptilian groups informally called *mammal-like-reptiles* and formally called the Synapsida. They evolved over an approximately 100 million-year period from the Pennsylvanian to the end of the Triassic period, when the first true mammals appeared. Some synapsids persisted into the Jurassic period, but by then the dinosaurs had proliferated. No other terrestrial tetrapods thrived until after dinosaurs went extinct at the end of the cretaceous period. The characters that can most clearly be reconstructed in fossils are those concerned with locomotion and feeding, because these are simply related to the form of preserved bones and teeth.

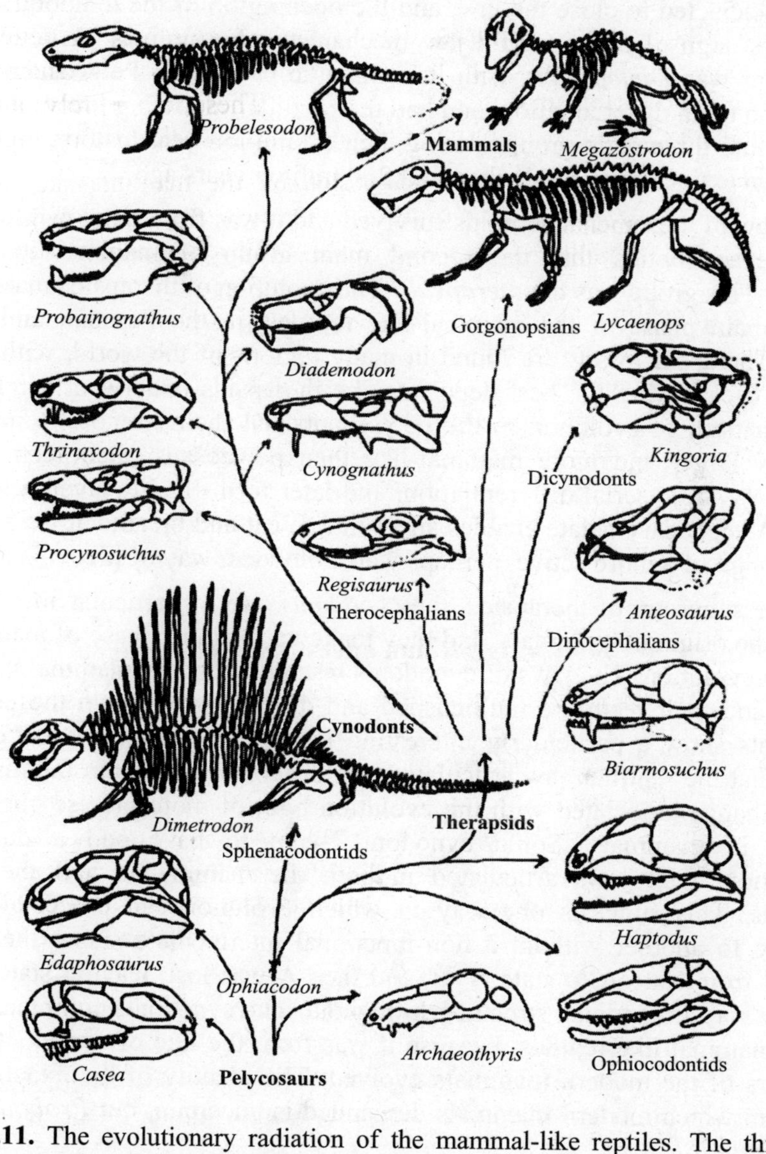

Figure 3.11. The evolutionary radiation of the mammal-like reptiles. The three main phases were: (a) pelycosaurs (sphenacodontids and ophiacodontids in this figure), (b) therapsids, and (c) cynodonts

We can distinguish three main phases in mammal-like reptilian evolution. The first phase corresponds to one of the two major divisions of the group, the *pelycosaurs*. Pelycosaur fossils are preserved from the Pennsylvanian and Permian, particularly from southwest USA where rocks of this age are located. *Archaeothyris* lived there about 300 million years ago, and is an early pelycosaur. It was a lizard-like animal, about 20 inches long. Its distinctive difference from other reptile groups is an opening in the bones behind the eye. The opening is called **temporal fenestra**, and in the living animal a muscle passed through it. The muscle acted to close the jaw, and the opening up of the temporal fenestrae is the first sign of the powerful jaw mechanism of mammals. A better known pelycosaur was *Dimetrodon,* with its enigmatic back-sails. Pelycosaurs showed little or no tooth differentiation, and had the reptilian sprawling locomotion. They evolved into three main groups during their 50 million-year history, and most of them went extinct quite suddenly about 260 million years ago.

A few of the sphenacodontids survived and it was from unknown line within the sphenacodontids that the second main group of mammal-like reptiles evolved. This group was the *therapsids.* The evolution of therapsids makes up the second main phase of the mammal-like reptiles, in the Permian and Triassic period. Therapsid fossils are found in many regions of the world, with a site in South Africa having the best deposits. The therapsids underwent a remarkably similar pattern of evolution to the pelycosaurs but their temporal fenestrae are generally larger and more mammal-like than pelycosaus, their teeth in some cases show more serial differentiation, and later forms had evolved a secondary palate. A secondary palate enables an animal to eat and breathe at the same time and is a sign of a more active, perhaps warm-blooded, way of life.

One subgroup of therapsids, the *cynodonts,* are of particular importance in tracing the origin of mammals, and they make up the third phase of mammal-like reptilian evolution. The jaws of cynodonts resemble modern mammal jaws more closely and their teeth are multicusped and differentiated down the jaw. Some cynodonts show a particularly interesting intermediate stage in jaw evolution. Recall that the reptilian jaw articulates in a different place from the mammalian jaw, a change associated with the evolution both of more precise chewing and hearing in mammals. Some cynodonts seem to have had a double jaw articulation; their jaws articulated in both the mammalian and the reptilian positions. This suggests one way in which evolution can proceed from one structure to another without a non-functional intermediate stage: the structure evolved from state A to state A+B, and then A was lost, leaving state B alone. The jaw was a functional structure throughout. The cynodonts complete the story of the mammal-like reptiles, because it was from the line of cynodonts that the ancestors of the modern mammals evolved. The identity of the exact cynodont line from which modern mammals descended is uncertain, but *Probainognathus* is closest to it.

3.5 EXTINCTION AND RADIATIONS

The diversity of life is affected by rates of loss and gain of new life forms. The loss of species is by extinction, and the gain of species is by speciation. When the speciation rate exceeds the extinction rate for a taxon, its diversity increases. When the extinction rate exceeds the speciation rate, its diversity decreases. Taxonomic groups often diversify extensively early in their history. In many cases, major adaptive radiations seem to have occurred after the evolution of some novel characteristic that opened a new adaptive zone, a set of new living conditions and resources that presents many previously unexploited opportunities. For example, the evolution of wings provided many new possibilities for insects, such as rapid access to treetops, islands, and other new feeding and breeding areas. Adaptive radiation can occur at all taxonomic levels, and on all geographic scales. For example, the proliferation of life on earth from the origin of life to the present is an adaptive radiation on the largest scale. However, adaptive radiations are particularly clear when they occur in a relatively small taxonomic group, in a confined geographic area. For example, Darwin's finches on the Galapagos Islands, where a single ancestral species has evolved into many species with a range of ecological adaptation.

Radiations occur in a number of circumstances

1. *Colonization of a new area where there are no competitors*. The radiation of Darwin's finches on island chains occurred when ancestral species colonized the respective area. Probably no competitors were present because the island had only recently emerged from the sea. The new area contains unexploited resources, and the ancestral species radiated into a range of forms that could exploit these resources.
2. *Extinction of competitors.* The radiation of mammals was followed by extinction of dinosaurs. The dinosaur extinction vacated the ecological space which was then occupied by mammals.
3. *Replacement of competitors.* One taxon may radiate if it is adaptively superior to its competitors. The superior will take over from the inferior one, which will be driven extinct. The superior taxon may have become superior because of environmental changes, or because it evolved a new, superior adaptation.
4. *Adaptive breakthroughs.* A taxon may evolve a new adaptation that allows it to compete another taxon, or may allow it to exploit a previously unexploited resource. For instance animals colonized the land as they evolved a set of appropriate adaptations.

The colonization of land was somewhat analogous to the colonization of a new island chain (point 1 above). However, we can distinguish between colonization of an existing area, made by the evolution of a new adaptation, and colonization of a new area that requires no new adaptation. The colonization of land required new adaptations for support, respiration, and water retention. By contrast, the finch that first colonizes the Galapagos Island probably did not have to evolve a new adaptation before it could colonize the island. After colonization of land, animals underwent extensive adaptive radiations and these radiations were made possible by adaptive breakthroughs.

3.5.1 Cambrian explosion: One of the most important adaptive radiations in the history of life

Based largely on anatomical and embryological criteria, animals are grouped into about 35 phyla. Animals in each phylum exhibit a distinctive combination of body features, a unique body plan, markedly different from that of other phyla. For instance, the basic features of the body plan of Phylum Arthopoda include jointed legs, an external skeleton, and segmentation.

The fossil record and molecular studies indicate that the diversification that produced the many animal phyla occurred rapidly on the vast scale of geological time. This relatively brief evolutionary episode probably lasted about 40 million years (about 565 to 525 million years ago) during the late Precambrian and early Cambrian (which began about 545 million years ago). Paleontologists have named the last period of the Precambrian era the *Ediacaran period*, for it was in the Ediacara hills of Australia, that fossils of Precambrian animals were first discovered. Similar animals of similar vintage have since been found on other continents. Most of the Ediacaran fossils (Figure 3.12) appear to represent cnidarians, but soft-bodied mollusks were also present, and numerous fossilized burrows and tracks indicate the activities of several groups of worms.

Figure 3.12. Ediacaran fossil animal: *Dickinsonia costata.*

Nearly all the major animal body plans appear in Cambrian rocks dating from 545 to 525 million years ago. During this relatively short span, a burst of animal origins called the Cambrian explosion left a rich fossil assemblage that includes the first animals with hard, mineralized skeletons. The Burgess Shale (Figure 3.13) in British Columbia, Canada, is the most famous fossil bed documenting the diversity of Cambrian animals. Two other fossil sites, one in Greenland and the other in the Yunnan region of China, predate Burgess Shale by more than 10 million years.

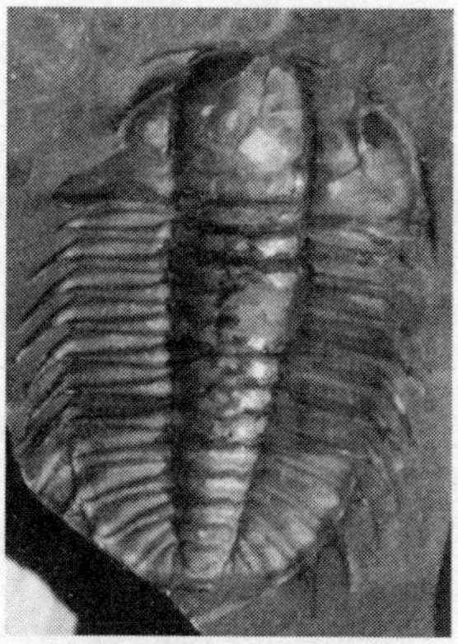

Figure 3.13 *Olenoides serratus*. This is the largest of several species of trilobites found in the Burgess Shale

Cambrian explosion is a fossil event, and probably marks the time of origin of hard parts. Animals with hard skeletons or shells leave fossils much more often than animals with only soft parts. But if hard parts originated about 540 million years ago, that raises the question of why hard parts suddenly became advantageous in so many groups at about the same time.

Predators in general is one hypothesis and visually hunting predators in particular is a second hypothesis. After predators evolved, hard parts became advantageous for defensive reasons. Another factor is that oxygen levels may have increased towards the end of the Precambrian era. This may have been caused by an increase in plant, that is phytoplankton- productivity. The higher plant production would have supported a greater mass and diversity of animals. The greater quantity of potential prey could have created an opportunity leading to the evolution of predators.

3.5.2 Mass extinctions

A species may become extinct because its habitat has been destroyed or because the environment has changed in a direction unfavorable to the species. If ocean temperature falls by even a few degrees, many species that are otherwise beautifully adapted will not survive. Even if physical factors in the environment are stable, biological factors may change, the environment in which a species lives includes the other organisms that live there, and evolutionary change in one species is likely to have some impact on other species in the community. For example, the evolution by some Cambrian animals of hard body parts, such as jaws and shells, may have made some organisms lacking hard parts more vulnerable to predation and thereby more prone to extinction.

Extinction is in fact inevitable in a changing world, and there have been crises in the history of life when global environmental changes have been so rapid and disruptive that a majority of species were swept away. Figure 3.14 shows the series of peak times when extinction rates appear to be exceptionally high. These peaks are called as *mass extinctions.* The exact definition of mass extinction is arbitrary, and different paleontologists recognize different number of mass extinctions in the history of life. From the Ordovician onwards, the five largest extinction events observed were in the late Ordovician, the late Devonian, the end of the Permian, the late Triassic, and the end of the Cretaceous period. These are sometimes called the "big five". Of these mass extinctions chronicled in the fossil record, two have received the most attention: the Permian extinctions is the biggest in history, with 80-96% of species going extinct. The Permian extinction occurred about 250 million years ago, and the Cretaceous extinctions, in this extinction at least half, and perhaps 60-75% of species went extinct. The Cretaceous extinction occurred about 65 million years ago.

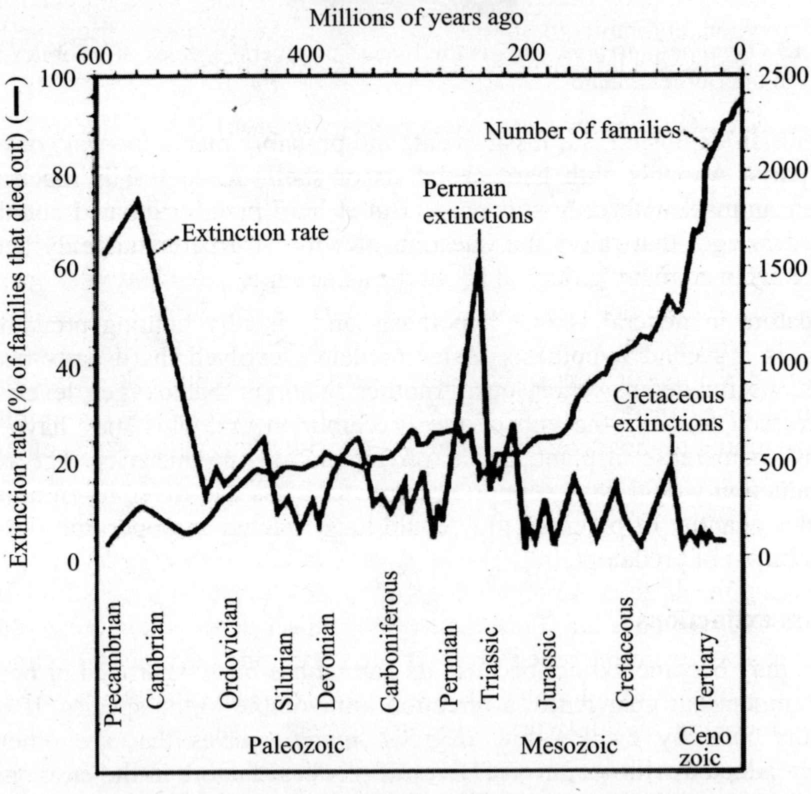

Figure 3.14. Periods of mass extinction and the diversity of life. Based on fossil records of terrestrial and marine organisms, this graph shows a general increase in the diversity of organisms with time, it also shows the mass extinction at various intervals which decreases the biological diversity. Of the many peaks showing extinction, the Permian and the Cretaceous mass extinctions are the major ones.

3.5.2.1 Permian extinction

The Permian extinctions, which define the boundary between the Paleozoic and Mesozoic eras, claimed about 90% of the species of the marine animals. Terrestrial life also crashed; for example, 8 out of 27 orders of Permian insects did not survive into the Triassic, the next geological period. This mass extinction occurred in less than 5 million years, possibly much less in the context of geological time. Several factors may have combined to cause radical environmental change during the late Permian. That was about the time the continents merged to form Pangaea, which disturbed many marine and terrestrial habitats and altered climate. Massive volcanic eruptions also occurred in what is now called Siberia, constituting the most extreme episode of volcanism of the past half-billion years. Besides eruption of lava and ash into the atmosphere, the Siberian volcanoes may have produced enough carbon dioxide to warm the climate. Reduced temperature differences between the equator and the poles may have slowed the mixing of ocean water, which in turn may have reduced the amounts of oxygen available to marine organisms. An oxygen deficit in the oceans may have played a large role in the Permian extinctions.

3.5.2.2 Extinction of dinosaurs (The Cretaceous extinction)

It was proposed in 1977 that Cretaceous extinction was caused by asteroid impact. Cretaceous-Tertiary boundary has high levels of iridium, rare in earth's crust but common in meteorites. Calculations of effects of nuclear bomb explosions ("nuclear winter") compare with worldwide climate cooling expected from large asteroid impact. Worldwide layer of soot also defines iridium layer. Huge meteorite crater of correct age found in Caribbean Ocean and Yucatan peninsula; suspected site of impact of meteor that resulted in dinosaur extinction, Marine animal fossil records indicate mass extinctions occur every 26 million years; corresponding to movement of solar system within the Milky Way galaxy.

The Alvarez Theory

Louis Alvarez, his son Walter, and their colleagues proposed that a giant asteroid or comet striking the earth some 65 million years ago caused the massive die-off at the end of the Cretaceous era. Presumably, the impact generated so much dust and gases that skies were darkened all over the earth, photosynthesis declined, and worldwide temperatures dropped. The outcome was that as many as 75% of all species — including all dinosaurs — became extinct.

The key piece of evidence for the Alvarez hypothesis was the finding of thin deposits of clay containing the element iridium at the interface between the rocks of the Cretaceous and those of the Tertiary period (called the K-T boundary after the German word for Cretaceous). Iridium is a rare element on earth (although often discharged from volcanoes), but occurs in certain meteorites at concentrations thousands of times greater than in the earth's crust.

After languishing for many years, the Alvarez theory gained strong support from the discovery in the 90s of the remains of a huge (180 km in diameter) crater in the Yucatan Peninsula that dated to 65 million years ago. The abundance of sulfate-containing rocks in the region suggests that the impact generated enormous amounts of sulfur dioxide (SO_2), which later returned to earth as a bath of acid rain. A smaller crater in Iowa, formed at the same time, may have contributed to the devastation. Perhaps during this period the earth passed through a swarm of asteroids or a comet and the repeated impacts made the earth uninhabitable for many creatures of the Mesozoic era.

3.5.2.3 Sixth mass extinction

There is little doubt left in the minds of professional biologists that earth is currently faced with a mounting loss of species that threatens to rival the five great mass extinctions of the geological past. As long ago as 1993, Harvard biologist E.O. Wilson estimated that earth is currently losing something of the order of 30,000 species per year — which breaks down to the even more discouraging statistic of some three species per hour. Some biologists have begun to feel that this biodiversity crisis — this "Sixth Extinction" — is even more severe, and more imminent, than Wilson had supposed. Can we stop the devastation of our planet and save our own species? We are in a biodiversity crisis — the fastest mass extinction in earth's history, largely due to: (1) human destruction of ecosystems, (2) overexploitation of species and natural resources, (3) human overpopulation, (4) the spread of agriculture, and (5) pollution.

At first glance, the physically caused extinction events of the past might seem to have little or nothing to tell us about the current Sixth Extinction, which is a patently human-caused event. And because *Homo sapiens* is clearly a species of animal (however behaviorally and ecologically peculiar an animal), the Sixth Extinction would seem to be the first recorded global extinction event that has a biotic, rather than a physical, cause.

3.6 COEVOLUTION

Some biologists have proposed that interactions between species also can result in microevolution in each of their populations. According to this hypothesis, when populations of two different species interact over a long time, changes in the gene pool of one species can lead to changes in the gene pool of the other species. This reciprocal microevolutionary change between populations of two different interacting species is known as *coevolution.* In coevolution each species exerts a selection pressure on, and evolves in response to the other species. The two lineages evolve together (Figure 3.15). In all examples of evolution, a species evolves in relation to changes in its environment. Coevolution refers to the special case of evolution in which the species environment is itself evolving.

Suppose that certain individuals in a population of carnivores (such as hawks) become better at hunting prey (such as mice). Because of genetic variation, certain individuals of prey have traits that allow them to escape or hide from their predators, and they pass these adaptive traits on to some of their offspring. However, a few individuals in the predator population also may have traits (such as better eyesight or quicker reflexes) that allow them to hunt the better-adapted prey successfully. They would then pass these traits on to some of their offsprings.

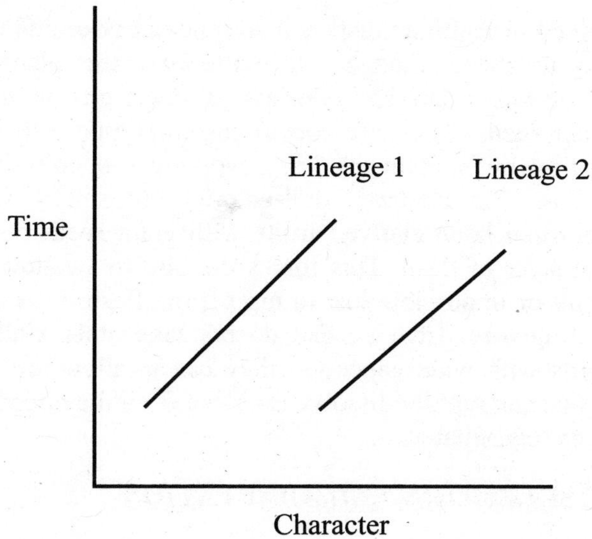

Figure 3.15. Coevolution means that two separate lineages mutually influence each other's evolution, and they evolve together.

3.6.1 Insect-plant coevolution

Insect-plant coevolution is one of the most studied example of coevolution. Ehrlich and Raven (1964) in their most influential modern discussion of coevolution, listed the food plants of the main butterfly taxa. Each family of butterflies feeds on a restricted range of plants, but these plants are in many cases not phylogenetically closely related. Ehrlich and Raven explained the diet patterns mainly in term of plant biochemistry. Plants produce natural insecticides-chemicals like alkaloids that can poison herbivorous (phytophagous) insects. Insects evolve resistance to these chemicals, for instance develops mechanism of detoxification. When a new detoxifying mechanism arises, it will open up a new array of food supplies, consisting of all those plants that produce the now harmless chemical. The insects can feed on them, and will diversify to exploit the resource. The result will be that each insect group can feed on a range of food plants; the range being set by the capabilities of the insect's detoxifying mechanisms. The range of food plants will form a biochemical group, but need not form a phylogenetic group because unrelated plants could use the same defensive chemicals. In turn, natural selection favours plants having improved insecticides. This biochemical events between plants and insects in which plants lineage develops new insecticides and in response to this the insects lineage develops new detoxifying mechanisms, might therefore promote the diversification of insects and angiosperms.

Another interesting example of coevolution is between fruit-eating birds and fruit-bearing plants. Many fruit-eating birds, especially in tropical rain forests, are coevolving with the plants whose fruits they eat. The birds get nourishment, and in the process the plants get their digestion-resistant seeds dispersed by regurgitation or along with the birds' droppings. Many characteristics of the plants have evolved to facilitate dispersal and the behavior and diets of the birds have responded to those changes. In particular, the plants have evolved conspicuously colored, relatively odorless fleshy fruits to attract the avian dispersers of their seeds. They are coevolving in response to the finely honed visual systems of the birds; plant species coevolving with color-blind mammalian seed-dispersers have, in contrast, dull-colored but smelly fruits. The bird-dispersed plants often have evolved fruits with giant seeds covered by a thin, highly nutritious layer of flesh. This forces the bird to swallow the fruit whole, since it is difficult or impossible just to nip off the flesh. In response, birds that are specialized frugivores (that is, that do not take other kinds of food) have evolved both bills with wide gapes (so they can swallow the fruit whole) and digestive tracts that can rapidly dissolve the flesh from the large impervious seed, which can then be regurgitated.

3.7 NATURAL SELECTION AND ADAPTATION

3.7.1 Adaptation

Adaptation is the evolutionary process whereby a population becomes better suited to its habitat. Adaptation is a heritable trait that enables organisms to better survive and reproduce under a given set of environment conditions. This process takes place over many generations, and is one of the basic phenomena of biology.

The term *adaptation* may also refer to a feature which is especially important for an organism's survival. For example, the adaptation of horses' teeth for the grinding of grass, or their ability to run fast and escape predators. Such adaptations are produced in a variable population by the better suited forms reproducing more successfully, that is, by natural selection.

Adaptation is, first of all, a *process*, rather than a physical part of a body. An internal parasite (such as a fluke) can illustrate the distinction: such a parasite may have a very simple bodily structure, but nevertheless the organism is highly adapted to its specific environment. From this we see that adaptation is not just a matter of visible traits: in such parasites critical adaptations take place in the life-cycle, which is often quite complex. However, as a practical term, adaptation is often used for the *product*: those features of a species which result from the process. Many aspects of an animal or plant can be correctly called adaptations, though there are always some features whose function is in doubt. By using the term *adaptation* for the evolutionary *process*, and *adaptive trait* for the bodily part or function (the product), the two senses of the word may be distinguished.

Adaptation is one of the two main processes that explain the diverse species we see in biology, such as the different species of Darwin's finches. The other is speciation (species-splitting or cladogenesis), caused by geographical isolation or some other mechanism. A favourite example used today to study the interplay of adaptation and speciation is the evolution of cichlid fish in African lakes, where the question of reproductive isolation is much more complex. In previous chapters we have met many more examples, such as camouflage in moths and drug resistance in HIV.

Adaptation is not always a simple matter, where the ideal phenotype evolves for a given external environment. An organism must be viable at all stages of its development and at all stages of its evolution. This places *constraints* on the evolution of development, behaviour and structure of organisms. The main constraint, over which there has been much debate, is the requirement that each genetic and phenotypic change during evolution should be relatively small, because developmental systems are so complex and interlinked. However, it is not clear what "relatively small" should mean, for example polyploidy in plants is a reasonably common large genetic change. The origin of the symbiosis of multiple micro-organisms to form a eukaryote is a more exotic example.

All adaptations help organisms survive in their ecological niches. These adaptive traits may be structural, behavioral or physiological. Structural adaptations are physical features of an organism (shape, body covering, armament; and also the internal organization). Behavioural adaptations are composed of inherited behaviour chains and/or the ability to learn: behaviours may be inherited in detail (instincts), or a tendency for learning may be inherited. Examples: searching for food, mating, vocalizations. Physiological adaptations permit the organism to perform special functions (for instance, making venom, secreting slime, phototropism); but also more general functions such as growth and development, temperature regulation, ionic balance and other aspects of homeostasis. Adaptation, therefore, affects all aspects of the life of an organism.

3.7.2 Types of adaptation

Changes in habitat

Before Darwin, adaptation was seen as a fixed relationship between an organism and its habitat. It was not appreciated that as the climate changed, so did the habitat; and as the habitat changed, so did the biota. Also, habitats are subject to changes in their biota: for example, invasions of species from other areas. The relative numbers of species in a given habitat are always changing. Change is the rule, though much depends on the speed and degree of the change.

When the habitat changes, three main things may happen to a resident population: habitat tracking, genetic change and extinction. In fact, all three things may occur in sequence. Of these three effects, only genetic change brings about adaptation.

Habitat tracking

When a habitat changes, the most common thing to happen is that the resident population moves to another locale which suits it; this is the typical response of flying insects or oceanic organisms, who have wide (though not unlimited) opportunity for movement. This common response is called *habitat tracking*. It is one explanation put forward for the periods of apparent stasis in the fossil record (the punctuated equilibrium thesis).

Genetic change

Genetic change is what occurs in a population when natural selection acts on the genetic variability of the population. By this means, the population adapts genetically to its circumstances. Genetic changes may result in visible structures, or may adjust physiological activity in a way that suits the changed habitat.

It is now clear that habitats and biota do frequently change. Therefore, it follows that the process of adaptation is never finally complete. Over time, it may happen that the environment changes little, and the species comes to fit its surroundings better and better. On the other hand, it may happen that changes in the environment occur relatively rapidly, and then the species becomes less and less well adapted. Seen like this, adaptation is a genetic *tracking process*, which goes on all the time to some extent, but especially when the population cannot or does not move to another less hostile area. Also, to a greater or lesser extent, the process affects every species in a particular ecosystem.

Van Valen thought that even in a stable environment, competing species had to constantly adapt to maintain their relative standing. This became known as the Red Queen hypothesis.

3.7.3 Coadaptation

Coadaptation here refers to complex adaptations, the evolution of which would have required mutually adjusted changes in more than one of their parts. One of the standard example of a complex coadaptation is the eye. When one eye part, such as the distance from the retina to the cornea, changes during evolution (Figure 3.16), changes in other parts, such as lens shape, would be needed at the same time. Because of the incompatibility of simultaneous correct mutations in both parts at the same time, a complex, finely adjusted engineering device like the eye could not therefore evolve by natural selection. The Darwinian reply is that the different parts could evolve independently in small steps; it is not necessary for all the parts of an eye to change at the same time in evolution.

In co-evolution, where the existence of one species is tightly bound up with the life of another species, new or 'improved' adaptations which occur in one species are often followed by the appearance and spread of corresponding features in the other species. There are many examples of this; the idea emphasizes that the life and death of living things is intimately connected, not just with the physical environment, but with the life of other species. These relationships are intrinsically dynamic, and may continue on a trajectory for millions of years, as has the relationship between flowering plants and insects (pollination).

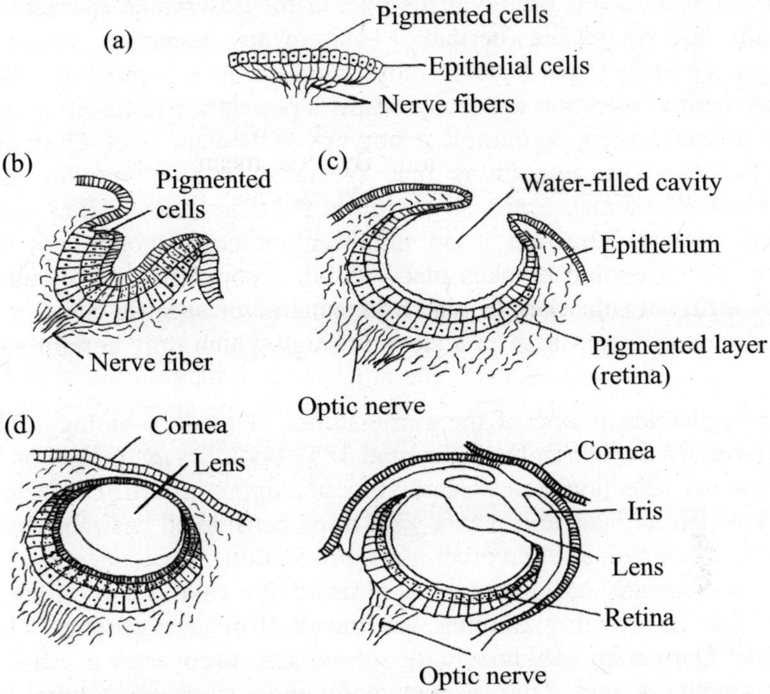

Figure 3.16. Stages in the evolution of eye, illustrated by species of mollucsks. (a) A simple spot of the pigmented cells. (b) A folded region of the pigmented cells, which increases the number of sensitive cells per unit area. (c) A pin-hole camera eye, as is found in *Nautilus*. (d) An eye is protected by adding a transparent cover of skin and part of the cellular fluid has differentiated into a lens and a full, complex eye, as found in the octopus and squid.

3.8 NATURAL SELECTION AS THE MECHANISM OF ADAPTIVE EVOLUTION

Natural selection is the process by which traits become more or less common in a population due to consistent effects upon the survival or reproduction of their bearers. It is a key mechanism of evolution.

The natural genetic variation within a population of organisms may cause some individuals to survive and reproduce more successfully than others in their current environment. For example, the peppered moth exists in both light and dark colours in the United Kingdom, but during the industrial revolution many of the trees on which the moths rested became blackened by soot, giving the dark-colored moths an advantage in hiding from predators. This gave dark-colored moths a better chance of surviving to produce dark-colored offspring, and in just a few generations the majority of the moths were dark. Factors which affect reproductive success are also important, an issue which Charles Darwin developed in his ideas on sexual selection.

Natural selection acts on the phenotype, or the observable characteristics of an organism, but the genetic (heritable) basis of any phenotype which gives a reproductive advantage will become more common in a population. Acting on phenotypes, natural selection indirectly adapts a population to its environment by increasing or maintaining favourable genotypes in the gene pool. Over time, this process can result in adaptations that specialize populations for particular ecological niches and may eventually result in the emergence of new species. In other words, natural selection is an important process (though not the only process) by which evolution takes place within a population of organisms. As opposed to artificial selection, in which humans favor specific traits, in natural selection the environment acts as a sieve through which only certain variations can pass.

Natural selection is one of the cornerstones of modern biology. The term was introduced by Darwin in his influential 1859 book *On the Origin of Species*, in which natural selection was described as analogous to artificial selection, a process by which animals and plants with traits considered desirable by human breeders are systematically favored for reproduction. The concept of natural selection was originally developed in the absence of a valid theory of heredity; at the time of Darwin's writing, nothing was known of modern genetics. The union of traditional Darwinian evolution with subsequent discoveries in classical and molecular genetics is termed the *modern evolutionary synthesis*. Natural selection remains the primary explanation for adaptive evolution.

The term *natural selection* has slightly different definitions in different contexts. It is most often defined to operate on heritable traits, because these are the traits that directly participate in evolution. However, natural selection is "blind" in the sense that changes in phenotype (physical and behavioral characteristics) can give a reproductive advantage regardless of whether or not the trait is heritable (non-heritable traits can be the result of environmental factors or the life experience of the organism).

Following Darwin's primary usage, the term is often used to refer to both the evolutionary consequence of blind selection and to its mechanisms. It is sometimes helpful to explicitly distinguish between selection's mechanisms and its effects; when this distinction is important, scientists define "natural selection" specifically as "those mechanisms that contribute to the selection of individuals that reproduce", without regard to whether the basis of the selection is heritable. This is sometimes referred to as "phenotypic natural selection".

Traits that cause greater reproductive success of an organism are said to be selected for, whereas those that reduce success are selected against. Selection for a trait may also result in the selection of other correlated traits that do not themselves directly influence reproductive advantage. This may occur as a result of pleiotropy or gene linkage.

3.8.1 Concept of fitness

The concept of fitness is central to natural selection. Broadly, individuals which are more "fit" have better potential for survival, as in the well-known phrase "survival of the fittest". However, as with natural selection above, the precise meaning of the term is much more subtle, and Richard Dawkins manages in his later books to avoid it entirely. (He devotes a chapter of his book, *The Extended Phenotype*, to discussing the various senses in which the term is used). Modern evolutionary theory defines fitness not by how long an organism lives, but by how successful it is at reproducing. If an organism lives half as long as others of its species, but has twice as many offspring surviving to adulthood, its genes will become more common in the adult population of the next generation.

Though natural selection acts on individuals, the effects of chance mean that fitness can only really be defined "on average" for the individuals within a population. The fitness of a particular genotype corresponds to the average effect on all individuals with that genotype. Very low-fitness genotypes cause their bearers to have few or no offspring on average; examples include many human genetic disorders like cystic fibrosis.

Since fitness is an averaged quantity, it is also possible that a favorable mutation arises in an individual that does not survive to adulthood for unrelated reasons. Fitness also depends crucially upon the environment. Conditions like sickle-cell anemia may have low fitness in the general human population, but because the sickle-cell trait confers immunity from malaria, it has high fitness value in populations which have high malaria infection rates.

3.8.2 The effect of selection on a varying characteristic can be stabilizing, directional, or diversifying

Natural selection can affect the frequency of a heritable trait in a population in three different ways, depending on which phenotypes in a varying population are favoured. These three modes of selection are called stabilizing selection, directional selection and diversifying selection. They can be depicted with graphs that show how the frequencies of different phenotypes change with time (Figure 3.17). This is most meaningful for quantitative traits that depend on many gene loci.

Stabilizing selection

It acts against extreme phenotypes and favours the more common intermediate variants. This mode of selection reduces variation and maintains the status for a particular phenotypic character. For example, stabilizing selection keeps the majority of human birth weights in the 3-4 kg range. For babies much smaller or larger than this, infant mortality is greater.

Directional selection

Directional selection tends to favor phenotypes at one extreme of the range of variation. Insecticide resistance is an example. DDT was a widely used insecticide. After a few years of extensive use, DDT lost its effectiveness on insects. Resistance to DDT is a genetic trait that the presence of DDT in the environment made into a favored trait. Only those insects resistant to DDT survived, leading over time to populations largely resistant to DDT. Another example is **Industrial melanism** in the peppered moth (Biston betularia)). Before the Industrial Revolution in the 18th and early 19th centuries, only light-colored moths were collected in light-colored woodlands in England. There was a rare, dark form. With the pollution caused by the burning of coal, the light-colored tree trunks became darker due to soot. The once rare dark-colored moths became more prevalent, while the once-common light-colored moths became increasingly rare. Reason: predation by birds. The color that had the greatest contrast with the background (tree trunk) was at a disadvantage. Cleanup of the forest during the 1950s caused the allele frequencies of light and dark moths to reverse to pre-Industrial Revolution levels, dark moths are now rare, light moths are now common. The resistance of many bacterial species to antibiotics is another example of directional selection. Over 200 species show some degree of antibiotic resistance, necessitating the development and more prudent use of a new generation of antibiotic medicines.

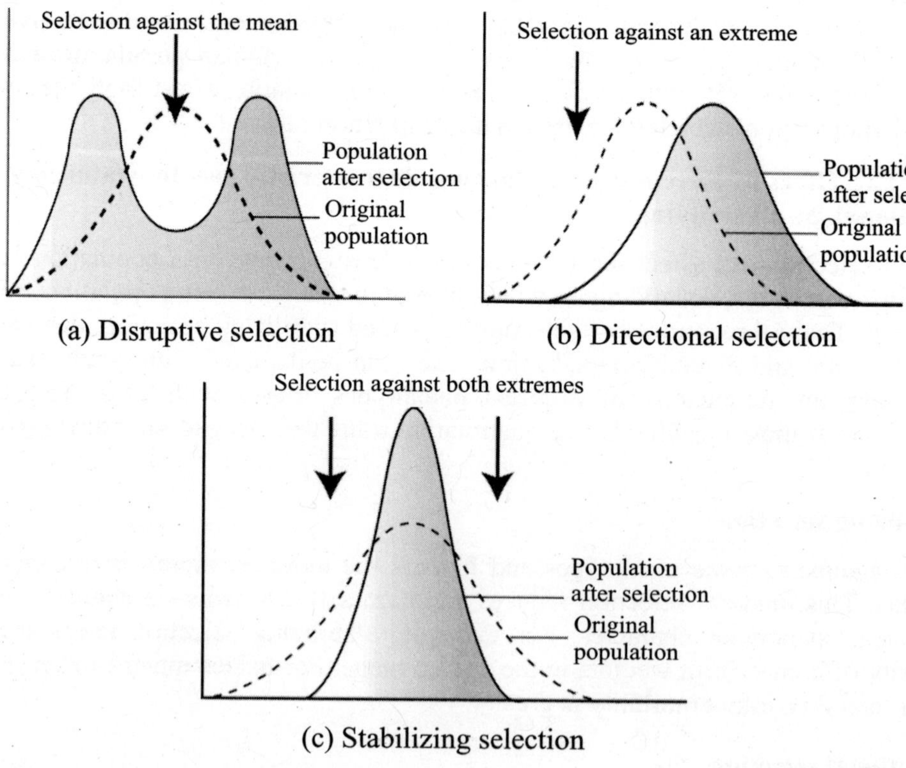

Figure 3.17. Modes of Selection.

Disruptive Selection

In some circumstances, individuals at both extremes of a range of phenotypes are favored over those in the middle. This is called disruptive selection. An example: The residues ("tailings") of mines often contain such high concentrations of toxic metals (e.g., copper, lead) that most plants are unable to grow on them. However, some hardy species (e.g. certain grasses) are able to spread from the surrounding uncontaminated soil onto such waste heaps. These plants develop resistance to the toxic metals while their ability to grow on uncontaminated soil decreases. Because grasses are wind-pollinated, breeding between the resistant and nonresistant populations goes on. But evidently, disruptive selection is at work. Higher death rates of both less resistant plants growing on contaminated soil and more resistant plants growing on uncontaminated soil leads to increasing divergence of the populations into two subpopulations with the extreme manifestations of this trait. The evolutionary significance of disruptive selection lies in the possibility that the gene pool may become split into two distinct gene pools. This may be a way in which new species are formed.

■■■■

Chapter 4

THE ORIGIN OF SPECIES

4.1 Species concept
- 4.1.1 The biological species concept
- 4.1.2 Recognition species concept
- 4.1.3 The phenetic species concept
- 4.1.4 The ecological species concept

4.2 Reproductive isolation barriers
- 4.2.1 Prezygotic Barriers
- 4.2.2 Postzygotic Barriers

4.3 Modes of speciation
- 4.3.1 Geographical barriers leading to Allopatric speciation
- 4.3.2 Adaptive radiation on island chains
- 4.3.3 Parapatric speciation
- 4.3.4 Sympatric speciation

4.4 Factors facilitating speciation

4.5 How much genetic change is required for speciation?
- 4.5.1 Punctuated equilibrium model

4.6 Polymophism

4.6.1 Balanced polymorphism

In the previous chapter we have described evolution at the smallest level, that is microevolution. It tends to explain evolution at genetic level of the population, but at the same time it is also important to explain the beginning of new forms of life–the origin of new species, which is at the focal point of evolutionary theory, for it is in new species that biological diversity arises. Evolutionary theory must also explain macroevolution, the origin of new taxonomic groups (new species, new genera, new families, even new kingdoms). **Speciation** is the keystone process because any genus, family, or higher taxon originates with a new species. The fossil record provides evidence for two patterns of speciation: anagenesis and cladogenesis.

1. **Anagenesis (phyletic evolution)** = The transformation of an unbranched lineage of organisms, sometimes to a state different

enough from the ancestral population to justify renaming it as a new species.

2. **Cladogenesis (branching evolution)** = The budding of one or more new species from a parent species that continues to exist; is more important than anagenesis in life's history, because it is more common and can promote biological diversity.

ANAGENESIS VS. CLADOGENESIS

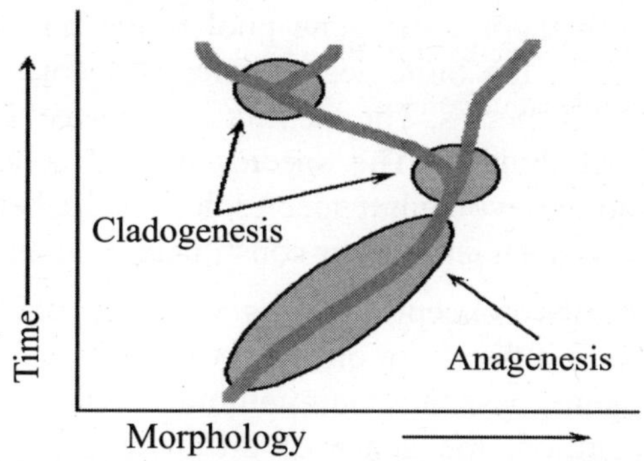

Before studying different types of speciation, it is important to know what actually a species is.

4.1 SPECIES CONCEPT

Species is a latin word meaning "kind" or "appearance." Indeed, we learn to distinguish between different kinds of organisms from differences in their appearance. Linnaeus, the founder of modern taxonomy, described individual species in terms of their physical form; the study of form or structure, called morphology or phenetic characters, is still the method most often used to characterize species. If one group of organisms consistently differs from other organisms, it will be defined as a separate species. The formal definition of the species will be in terms of characters that can be used to recognize members of that species. The taxonomist who describes the species will have examined specimens of it and of related species, looking for characters that are present in specimens of the species to be described, and absent from other closely related species. These are the characters used to define the species. Modern taxonomists also consider differences in body function, biochemistry, behaviour, and genetic makeup. Here we will discuss some commonly used species concept to describe species.

4.1.1 The biological species concept

The biological species concept is widely accepted species definition was first described by evolutionary biologist Ernst Mayr in 1942. Mayr's definition is answer to the question, what factors divide biological diversity into separate forms that we identify as species?

According to this concept, "species are a group of population which can freely interbreed in nature and produces viable, fertile offsprings and are ''reproductively isolated'' from members of other species." The expression reproductively isolated means that members of the species do not interbreed with members of other species, because they have some attributes that prevent interbreeding. The biological species concept is important because it places the taxonomy of natural species within the conceptual scheme of population genetics. In terms of population genetics, a group of interbreeding organisms make up a gene pool. In theory, the gene pool is the unit within which gene frequencies can change. In the biological species concept, gene pools become more or less identifiable as species. The identity is imperfect because species and populations are often subdivided. The species in this concept is the unit of evolution. Organisms do not evolve but species do, and higher taxonomic groups such as phyla only evolve in so far as their constituent species are evolving.

The biological species concept explains why the members of a species resemble one another, and differ from other species. When two organisms breed within a species, their genes pass into their combined offspring, as this process is repeated in every generation, the genes of different organisms are constantly shuffled around the species gene pool. Different family lineages (parents, offspring, grandchildren, and so on) soon become blurred by the transfer of genes between them. The shared gene pool gives the species its identity. By contrast, genes are not transferred to other species, and different species therefore evolve a different appearance. The movement of genes through a species by migration and interbreeding is called gene flow. According to the biological species concept, gene flow explains why each species forms a phenetic cluster.

In summary, the biological species concept is based on reproductive isolation, with each species isolated by factors (barriers) that prevent interbreeding, thereby blocking genetic mixing with other species.

Limitation of biological species concept

The biological species concept does not work in all situations. For instance, it is inadequate for grouping extinct forms of life, the fossils of which must be classified according to morphology. The criterion of interbreeding is meaningless for organisms that are completely asexual in their reproduction, as are all prokaryotes, some protists, some fungi, and even some plants (such as commercial banana). Many bacteria do transfer genes on a limited scale by conjugation and other processes, but there is nothing similar to the equal contribution of genetic material from two parents that occur in sexual reproduction. Different lineages of descent give rise to clones, which genetically speaking, represent single individuals. Asexual organisms can be assigned to species only by the grouping of clones that have the same structural and biochemical characteristics.

4.1.2 Recognition species concept

In 1993, Paterson gives the recognition species concept, which is closely related to the biological species concept. Paterson defines a species as a set of organisms with a shared mate recognition system (SMRS). This system helps an individual to recognize a potential mate on the basis of the morphological, behavioural characteristics. For example, as many as 30 or 40 different species of crickets (they are insects belonging to the family Gryllidae, and are related to grasshoppers) may be breeding within a single habitat in USA (Figure 4.1). The male crickets broadcast their songs and are approached by females. Interbreeding is confined within a species because each species has its own distinctive song and females only approach males that are singing their species song. The system of a male song and a female auditory system that leads females to approach some songs and not others is an example of SMRS. The set of organisms that are defined as a species by the biological and recognition species concepts will be very similar, because organisms that interbreed will usually also have a shared SMRS. However, both biological and recognition species concept focus on mating, so their implementation is limited only to sexually reproducing organisms (particularly animals).

Figure 4.1. An American cricket *Gryllus bimaculatus*

4.1.3 The phenetic species concept

This species concept can be understood as an extension of the way taxonomists define species. Taxonomists define each species by a particular defining character, or characters, that are shared by its members. In general, it defines a species as a set of organisms that are phenetically similar, and distinct from other set of organisms.

The classical version of the phenetic species concept is the "typological species concept" (the term morphological species concept has also been used to refer to the same concept). The word "typological" comes from the word "type", which is commonly used in taxonomy. When a new species is named, its description is based on a specimen called the type specimen, which has to be deposited in a public collection. According to the typological species concept, a species consists of all individuals that look sufficiently similar to the type specimen of the species.

A later version of the phenetic species concept was developed by the school of numerical taxonomists in the 1960s. They developed statistical techniques for describing the phenetic similarity of organisms. Those techniques could be applied to recognize species. A species could then be defined as a set of organisms of sufficient phenetic distinctness

4.1.4 The ecological species concept

The ecological species concept defines species on the basis of where they live and what they do instead of what they look like. The form and behaviour of organisms of a species, to some extent, adapted to the resources they exploit and the habitats they occupy. According to ecological researches, particularly with closely related species living in the same area, has abundantly demonstrated that the differences between species in form and behaviour are often related to differences in the ecological resources they exploit. The set of resources and habitats exploited by the members of a species form that species ecological niche and the ecological species concept defines a species as the set of organisms exploiting a single niche. In some cases, the full definition can be elaborated. If, for instance, the juvenile stage of an organism lives in plankton while the adult stage is attached to rocks, then the different life stages exploit different ecological niches. However, the definition could be expanded to define a species as a set of organisms who exploit a certain set of niches, where the set includes the niches exploited by different life stages, genders, or other forms within the species.

Species are present in discrete form; this can be explained by ecological processes. Consider a parasite exploiting resources from two host species. The host species will differ from each other in some respects, perhaps in where they live, or the times of day they are active, or their morphology. The parasites will evolve apropriate adaptations to live in one or the other host. The parasites then tend to become two discrete species, because their environmental resources (host in this case) come in two discrete kinds.

Two host species can clearly provide two discrete sets of ecological resources. But in other cases, the ecological resources may not come in such discrete units. Consider, for example, the five species of warblers in Maine that were the subject of a classic study by MacArthur (1958). MacArthur showed that each species mainly exploited a particular subregion of the trees they all lived in. Some species foraged higher, some lower; some foraged near the ends of branches, others nearer the center of the tree. These variables, like height in tree, are continuous. The warblers form five distinct, discrete species, but they divide up resource variables that are continuous. In this case, an ecological explanation for the existence of discrete species mainly comes from Gause's competitive exclusion principle. Only species that are sufficiently different can coexist. The result is that even with a continuous resource distribution, species may evolve into a series of discrete forms along the continuum. If the species blurred into one another, superior competitors could drive inferior competitors extinct, and gaps between species would appear.

The ecological and biological species concepts are closely related. Life, according to ecological species concept, comes in the form of discrete species because of adaptation to exploit the resources in nature. Interbreeding is shaped by the same process. Natural selection will favour organisms who interbreed with other organisms that have a similar set of ecological adaptations. For instance, the ecological adaptation might be the size of the beak, if the beak is adapted to eat seeds found locally. Natural selection favours individual birds that interbreed with other birds that have similar beaks. Then they will on average produce offsprings that are well adapted to eat the local seeds. Natural selection works against birds that interbreed with mates that have very different beaks as their offspring will tend to have maladapted beaks. The pattern of interbreeding and the ecological adaptations in a population are therefore shaped by common evolutionary forces.

4.2 REPRODUCTIVE ISOLATION BARRIERS

Any barrier or factor that impedes two species from producing viable, fertile hybrids contributes to reproductive isolation. No single barrier may be completely impenetrable to genetic exchange, but many species are genetically isolated by more than one type of barriers. Here we are considering only biological barriers to reproduction, which are intrinsic to the organisms. Of course, if two species are geographically segregated, they cannot possibly interbreed, but a geographical barrier is not considered equivalent to reproductive isolation because it is not intrinsic to the organism themselves. Intrinsic reproductive barriers prevent population belonging to different species from interbreeding, even if their ranges overlap. Intrinsic reproductive barriers are actually "evolved characters" that prevents interbreeding between different species. For instance, courtship is an example of reproductive barrier. If two species do not interbreed because their courtship differs, then the courtship behaviour of at least one of those species must have undergone evolutionary change and it became a new evolved character which leads to reproductive isolation.

The various types of reproductive barrier that isolate the gene pools of species can be categorized as prezygotic or postzygotic, depending on whether they function before or after the formation of zygotes or fertilized eggs.

4.2.1 Prezygotic Barriers

Prezygotic barriers prevent mating between species or impede formation of hybrid zygotes.

1. Habitat isolation

Two species that live in different habitats within the same area may encounter each other rarely, if at all, even though they are not technically geographically isolated. For example, two species of garter snakes in the genus *Thamnophis* occur in the same areas, but one lives mainly in water and the other is primarily terrestrial. Habitat isolation also affects parasites, which are generally confined to certain plant or animal host species. Two species of parasites living on different hosts will not have a chance to mate.

2. Behavioural isolation

It is also commonly known as ethological isolation, it represents special signals that attract mates, as well as elaborate behaviour unique to a species. It is probably the most important reproductive barrier among closely related animals. Recognition species concept is based on this type of reproductive barrier. Male fireflies of various species signal to females of their kind by blinking their lights in particular patterns. The females respond only to signals characteristic of their own species, flashing back and attracting the males.

The occurrence of different shape, colouration, habitat and the range of overlapping regions may be some of the deciding factors in behavioural isolation. The courtship rituals are specific to a species and these courtship rituals are another form of behavioural isolation. The distinctive songs of many birds, the special mating calls of certain frogs, and the sexual displays of most animals are generally attractive only to mates of the same species. Many plants have floral displays that attract only certain insect pollinators. Even where the morphological differences between two species are minimal, behavioural differences may prevent cross-fertilization. Thus *Drosophila melanogaster* and *Drosophila simulans,* designated as sibling species because of their morphological similarity, normally do not mate with each other even when kept together in a single population cage. According to Coyne and co-workers (1994), male courtship in this group depends on their attraction to specific hydrocarbons in the female cuticle, and this behavioural isolation can be caused by only few genetic differences.

3. Temporal isolation

The time period of mating of different species are never always same; two species that breed during different times of the day, different seasons, or different years cannot mix their gametes. This differential time period of mating which prevents interbreeding between two different species is known as temporal isolation. Three species of the orchid genus *Dendrobium* living in the same rain forest does not hybridize because they flower on different days. Pollination of each species is limited to a single day because the flowers open in the morning and wither that evening.

4. Mechanical isolation

Closely related species may attempt to mate, but cannot achieve the act or fail to achieve fertilization because their genetilia are anatomically not compatible or incompatible. This type of incompatibility, long thought to be a primary isolating mechanism in animals, even such kind of barriers contribute to reproductive isolation of flowering plants that are pollinated by insects or other animals. Floral anatomy is often adapted to certain pollinators that transfer pollen only among plants of the same species.

5. Gametic isolation

Even if the gametes of different species meet, they rarely fuse to form a zygote. For animals whose eggs are fertilized within the female reproductive tract (internal fertilization), the sperm of one species may not be able to survive in the environment of the female reproductive tract of another species. Many aquatic animals release their gametes into the surrounding water, where the eggs are fertilized (external fertilization). Even when two closely related species release their gametes at the same time in the same place, cross specific fertilization is uncommon. Gametes recognition may be based on the presence of specific molecules on the coats around the egg, which adhere only to complementary molecules on sperm cells of the same species. In some *Drosophila* crosses, Patterson and Stone and others have shown that an insemination reaction in the vagina of the female causes swelling and prevents successful fertilization of the egg. A similar mechanism of molecular recognition enables prevents growth of pollen tube in the styles of foreign species.

4.2.2 Postzygotic Barriers

If the prezygotic barriers do not impede the successful fertilization of the gametes belonging to different species, then postzygotic barriers prevent the hybrid zygote from developing into a viable, fertile adult.

1. Reduced Hybrid Viability or Hybrid Inviability

When prezygotic barriers are crossed and hybrid zygotes are formed, genetic incompatibility between the two species may abort development of the hybrid at some embryonic stage. Of the numerous species of frogs belonging to the genus *Rana*, some live in the same regions and habitats, where they may occasionally hybridize. But the hybrids generally do not complete their development.

2. Reduced Hybrid Fertility or Hybrid Sterility

Even if two species mate and produce the hybrid offspring that are vigorous, reproductive isolation is intact if the hybrids are completely or largely sterile. Since the infertile hybrids cannot backbreed with either parental species, genes cannot flow freely between the species. One cause of this barrier is a failure of the meiosis to produce normal gametes in the hybrid if chromosomes of the two parent species differ in number or structure. A familiar case of a sterile hybrid is the mule, a robust cross between a horse and a donkey; horses and donkeys remain distinct species because, except very rarely, mules cannot backbreed with either species (Figure 4.2).

Figure 4.2. Mule

3. Hybrid breakdown

In some cases when species crossmate, the first-generation hybrids are viable and fertile, but when these hybrids mate with one another or with either parent species, offsprings of the next generation are feeble or sterile. For example, different cotton species can produce fertile hybrids (F_1 generation), but breakdown occurs in the next generation (F_2 generation) when offspring of the hybrids die as seeds or grow into weak and defective plants

4.3 MODES OF SPECIATION

The crucial event for the origin of a new species is reproductive isolation. The members of a species usually differ genetically, ecologically, and in their behaviour and morphology from other species, as well as in who they will interbreed with. Some biologists prefer to define species not by reproductive isolation but by other properties, such as genetic or ecological differences. Probably no single property can provide a universal species definition, applicable to all animals, plants and microorganisms. However, many species do differ by being reproductively isolated, and even if the evolution of reproductive isolation is not always the crucial event in speciation, it is certainly the key event in research on speciation.

In 1889 A. R. Wallace proposed that natural selection might favour the establishment of mating barriers among population if the hybrids were adaptively inferior. That is, genotypes that did not mate to produce inferior hybrids would be selected over genotypes that did. According to this hypothesis, which Dobzhansky and others supported, selection for reproductive isolation arises because most races and species are strongly adapted to specific environments. Hybrids between two such highly adapted populations therefore represent a genetic dilution of their parental gene complexes that can be of great disadvantage in the original environments. Thus genotypes that have incorporated prezygotic isolating mechanisms would have the advantage of not wasting their gametes in producing deleterious offspring.

There are three general modes of speciation based on different geographic conditions and how the gene flow is interrupted in the speciating populations. If a new species evolves in geographic isolation from its ancestors, the process is called *allopatric speciation.* If the new species evolves in a geographically contiguous population, it is called *parapatric speciation.* If the new species evolves within the geographic range of its ancestor, it is called as *sympatric speciation.*

4.3.1 Geographical barriers leading to Allopatric speciation

Geological processes can fragment a population into two or more isolated populations. A mountain range may emerge and gradually split a population of organisms that can inhabit only lowlands; a land bridge, such as the Isthmus of Panama, may form and separate the marine life on either side; or a large lake may subside until there are several smaller lakes with their populations now isolated. If individuals colonize a new, geographically isolated area, the colonizing population may become isolated from the parent population.

Just how formidable a geographical barrier must be to keep allopatric populations apart depends on the organism's ability to disperse, i.e. on the mobility of animals or the dispersibility of spores, pollen, and seeds of plants. The Grand Canyon is a steep-sided canyon carved by the Colorado river in the United State in the state of Arizona is easily crossed by hawks and many other birds, but it is an impassable barrier to populations of small rodents confined to either north or south rim of the canyon. Indeed the same bird species populate the both rims of the canyon, but each rim has several unique species of rodents (Figure 4.3).

Lets consider an example how geographical barrier can lead to allopatric speciation. About 50,000 years ago, during an ice age, what is now the Death Valley region of California and Nevada had a very rainy climate and a system of interconnected lakes and rivers. A drying trend began about 10,000 years ago, and by 4,000 years ago the region had become a desert. Today, all that is left of the network of lakes and rivers is isolated springs scattered in the desert, mostly in deep clefts between rocky walls. The springs vary extensively in water temperature and salinity. Living in many of the springs are tiny fishes, called pupfishes of the genus *Cyprinodon*. Each inhabited spring, often no more than a few meters in diameter, is home to its own species of pupfish adapted to that pool and found nowhere else in the world. The various pupfishes probably descended from a single ancestral species whose range was broken up when the region became arid, cloistering several small populations that subsequently diverged from one another.

The Grand Canyon

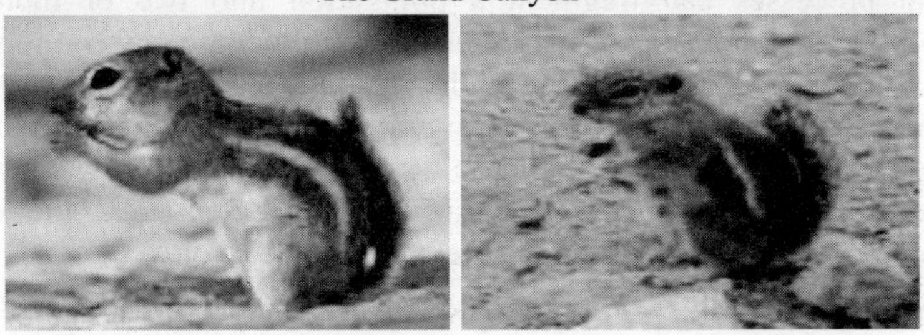

(a) *Ammospermophilus leucurus* (b) *Ammospermophilus harrisi*

Figure 4.3. Allopatric speciation of squirrels in the Grand Canyon. One example of allopatric speciation is the two species of antelope squirrels that inhabit opposite rims of the Grand Canyon. On the south rim is Harris's antelope squirrel (*Ammospermophilus harrisi*). A few miles away on the north rim is the closely related white-tailed antelope squirrel (*Ammospermophilus leucurus*), which is slightly smaller and has a shorter tail that is white underneath. Birds and other organisms that can disperse easily across the canyon have not diverged into different species on opposite rims.

Favourable conditions for allopatric speciation

Whenever population becomes allopatric, it is possible for speciation to occur as the isolated gene pools accumulate genetic differences by microevolution. But an isolated population that is small is more likely than a large population to change substantially enough to become a new species.

The geographical isolation of a small population usually occurs at the fringe of the parent populations range. The splinter population, or *peripheral isolate***,** is a good candidate for speciation for three reasons:

1. The gene pool of the peripheral isolate probably differs from that of the parent population from the outset. Living near the border of the range, the peripheral isolate represents the extremes of any genotypic clines that existed in the original population. And if the peripheral

isolate is small, there will be a founder effect resulting in a gene pool that is not representative of the gene pool of the parent population.

2. Until the peripheral isolate becomes a large population, genetic drift will continue to change its gene pool at random. New mutations or combination of existing alleles that are neutral in adaptive value may become fixed in the population by chance alone, causing genotypic and phenotypic divergence from the parent population.
3. Evolution caused by natural selection may take a different direction in the peripheral isolate than in the parent population. Because the peripheral isolate inhabits the frontier, where the environment is somewhat different, the peripheral isolate will probably encounter selection factors that are different from, and generally more severe than, those operating on the parent population.

These factors may cause peripheral isolate to follow an evolutionary course that diverges from that of parent population, as long as the gene pools remain isolated. This does not mean that all peripheral isolates persists long enough or change enough to become a new species. Life on the frontier is usually harsh, and most pioneer populations probably become extinct. As evolutionary biologist Stephan Jay Gould puts it: "Status as a peripheral isolate merely gives a lottery ticket to a small population. A population can't win (speciate) without a ticket, but there are very few winning tickets."

4.3.2 Adaptive Radiation on Island Chains

Islands are living laboratories for the study of speciation. Flurries of allopatric speciation have occurred on island chains where organisms that have strayed or become passively dispersed from their parent populations have founded new populations that evolved in isolation. The many indigenous species of the Galapagos descended from stragglers that floated, flew, or were blown over the sea from the South American mainland. For example, consider the Galapagos finches. A single dispersal event may have seeded one island with a small population of the ancestral finch, and this peripheral isolate formed a new species. Later, a few individuals of this island species may have reached neighbouring islands, where geographical isolation permitted additional speciation episodes. After diverging on one of these other islands, a new species could recolonize the island from which its founding population emigrated and coexist there with its parent species, or form still another species. Multiple invasions of islands by peripheral isolates of species from neighbouring islands would eventually lead to coexistence of several species on each island. The islands are far enough apart to permit populations to evolve in isolation, but close enough together for occasional dispersion events to occur. The evolution of many diversely adapted species from a common ancestor is called **adaptive radiation** (Figure 4.4). Adaptive radiation of the Galapagos finches is evident in the many types of bills specialized for different tools (see Figure 2.5).

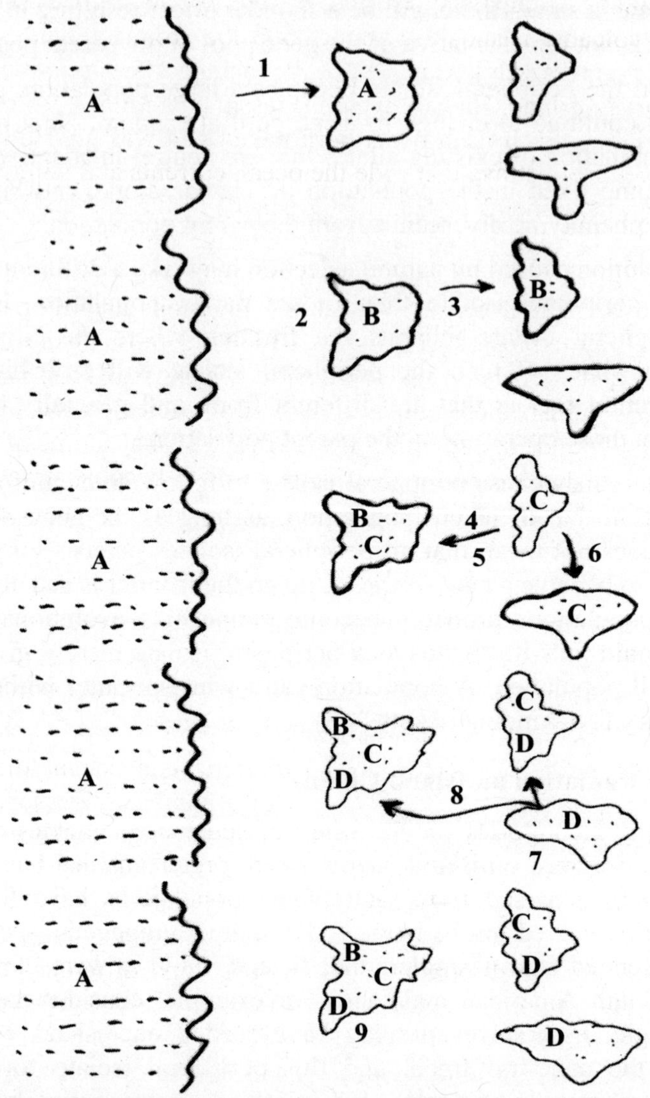

Figure 4.4. A model for adaptive radiation on island chains. 1. One island in this cluster of three is seeded by a small colony founded by members of species A, blown over from a mainland population. 2. Its gene pool now isolated from the parent species, the island population evolves into species B as it adapts to its new environment. 3. Storms or other agents of dispersion spread species B to a second island, 4. Where the isolated colony evolves into species C. 5. Later, individuals from species C recolonize the first island and cohabit with species B, but reproductive barriers keeps the species distinct. 6. A colony of species C may also populate a third island, 7 where it adapts and forms species D. 8. Species D is dispersed to the two islands of its ancestors, 9 and evolves into a new species, E, on one of those islands. This process would go on, with a series of allopatric speciation episodes made possible by the combination of isolation and occasional dispersal.

The Hawaiian Archipelago is one of the world's great showcases of evolution. The volcanic islands are about 3500 km from the nearest continent. They become progressively younger to the southeast, terminating with the youngest and largest island, Hawaii, which is less than a million year old and still has active volcanoes. Each island was born naked and was gradually populated by species derived from strays that rode the ocean currents and winds either from distant islands and continents or from older islands of the archipelago itself. The physical diversity of each island, including a range of altitudes and extensive differences in rainfall, provides many environmental opportunities for evolutionary divergence by natural selection. Multiple invasions and allopatric speciations have ignited an explosion of adaptive radiation; most of the thousands of species of plants and animals that now inhabit the islands are found nowhere else in the world. In contrast, there are no indigenous species on the Florida Keys. Apparently, those islands are so close to the mainland that founding populations are not sufficiently sequestered for their gene pools to become isolated from the steady stream of immigrants from the parent population on the mainland.

4.3.3 Parapatric speciation

In parapatric speciation, the new species evolve from contiguous populations, rather than completely separate ones, as in allopatric speciation. The full process could occur as follows. Initially, one species is distributed in space. The species evolves a "stepped cline pattern of geographical variation."

A cline is a gradient of continuous variation, in a phenotypic or genetic character, within a species. Clines can arise for a number of reasons. As we go towards north there is gradual increase in the body size of endotherm species (for example house sparrows), this is because the body size may be adapted to environmental temperature. Temperature gradually decreases to the north and body size in the sparrow's increases as we go north. Alternatively, the environment may change discontinuously in space and different genes may be adapted to the two regions. A cline can then arise because of gene flow: the movements of individuals, or their pollen in case of plants. Clines may be smooth or "stepped" depending on how suddenly gene frequencies change in space. If the environment changes smoothly, the cline will also be smooth. If the environment changes more suddenly, the cline may be more stepped. The shape of the step depends on the fitness difference between the genotypes in the two regions, the fitness of the intermediate genotypes (such as heterozygotes or recombinants), and the amount of gene flow. A sudden change in the environment is called "ecotone". However, ecotones are not the only explanation for stepped clines. Stepped clines may also result when the ranges of the two formerly separated populations expand and the two populations meet up or they may result from genetic drift. The stepped cline could exist because of an abrupt environmental change: one form of the species would be adapted to the conditions on one side of the boundary, the other form to the conditions on the other side of the boundary.

A *hybrid zone* is a stepped cline where two related populations that diverged after becoming geographically isolated make secondary contacts and interbreed where their geographical range overlaps. They are sufficiently different to be easily recognized. The two forms may have been given different names, as subspecies or races, or they may be different enough to have been classified as separate species.

The carrion crow (*Corvus corone*) and the hooded crow (*Corvus cornix*) in Europe are a classic example of species round a hybrid zone (Figure 4.5). The hooded crow is distributed more to the east, the carrion crow to the west, with the two species meeting along a line in central Europe. At the line i.e. the hybrid zone, they interbreed and produce hybrids. The hybrid zone for the crows was first recognized phenotypically, because the hooded crow is gray with a black head and tail, whereas the carrion crow is black all over. The two species are now known to differ in many other respects too. The fact that the crows interbreed in the hybrid zone means that speciation between them is incomplete.

The conditions in a hybrid zone (or a stepped cline) are particularly ripe for speciation if it is a *tension zone*. A tension zone exists when the hybrids between the forms on either side of the boundary are selectively disadvantageous. (A hybrid zone is not a tension zone if the hybrids have intermediate, or superior, fitness to the pure forms). For instance, if one homozygote (AA) is adapted to one environment, and another homozygote (aa) to another environment, heterozygotes (Aa) will be produced where the two environment meet up. If the heterozygotes are disadvantageous, the meeting place is an example of a tension zone.

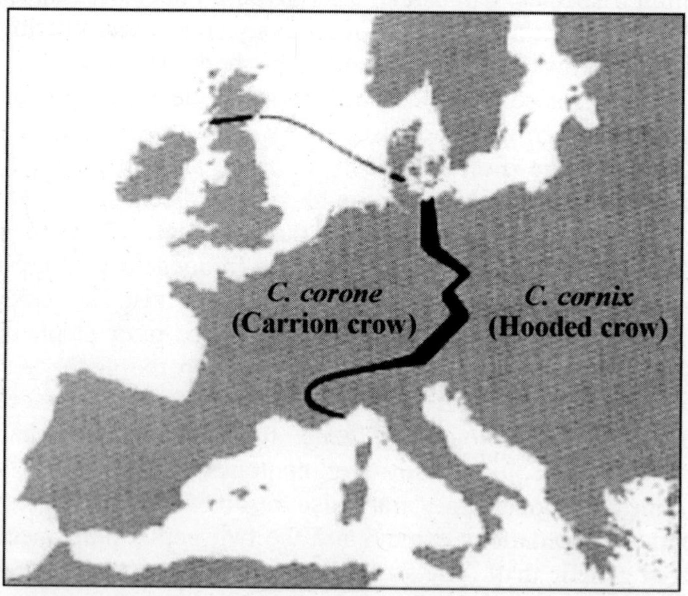

Figure 4.5. Hybrid zone between the carrion crow (*Corvus corone*) and hooded crow (*C. cornix*) in Europe. Here the two crows are shown as two separate species, but taxonomist classify them as subspecies.

In a tension zone, the conditions are exactly the preconditions for reinforcement. Matings within a type are advantageous, and mating between types produce disadvantageous hybrids. Natural selection favours assortative mating. We can therefore imagine a sequence where a stepped cline initially evolves, and then becomes distinct enough to count as a hybrid zone. We are near the border of the origin of a new species. Reinforcement could then finish speciation off, eliminating hybridization from the hybrid zone. That sequence of events constitutes parapatric speciation.

4.3.4 Sympatric speciation

In sympatric speciation, new species arise within the range of parent populations; genetic isolation evolves in a variety of ways without geographical isolation. For instance, in a single generation a new species (defined by the biological species concept) can be generated if a genetic change results in a reproductive barrier between the mutants and the parent population. Many plant species have their origins in accidents during the cell division that result in extra sets of chromosomes, a mutant condition called **polyploidy**. An **autopolyploid** is an individual that has more than two chromosome sets, all derived from a single species. For example, a failure of meiosis during gamete production can double chromosome number from the diploid count (2n) to a tetraploid number (4n) (Figure 4.6a). The tetraploid can then fertilize itself (self-pollinate) or mate with other tetraploids. However, the mutants cannot interbreed successfully with diploid plants of the original population; the offspring, which would be triploid (3n), would be sterile because unpaired chromosomes result in abnormal meiosis. In just one generation, a postzygotic barrier has caused reproductive isolation and has interrupted gene flow between a tiny population of tetraploids (may be only a single plant initially) and the parent diploid population that surrounds it.

Sympatric speciation by autopolyploidy was first discovered in early 1900s by the geneticist Hugo de Vries while he was studing the genetics of the evening primrose, *Oenothera lamarckiana*, a diploid species with 14 chromosomes. One day, de Vries noticed an unusual variant that had appeared among his plants, and microscopic inspection revealed that it was a tetraploid with 28 chromosomes. He found that the plant was unable to breed with the diploid primrose, and he named the new species *Oenothera gigas*.

Another type of polyploidy species, much more common than autopolyploids, is called an **allopolyploid**, referring to the contribution of two different species to a polyploidy hybrid (Figure 4.6b). The origin of an allopolyploid begins when two different species interbreed and combine their chromosomes. Interspecific hybrids are usually sterile because the haploid set of chromosomes from one species cannot pair during meiosis with the haploid set from the other species. Though infertile, a hybrid may actually more vigorous than its parent and propagate itself asexually (which many plants can do). At least two mechanisms, illustrated in Figure 4.6b, can transform the sterile hybrids into fertile polyploids.

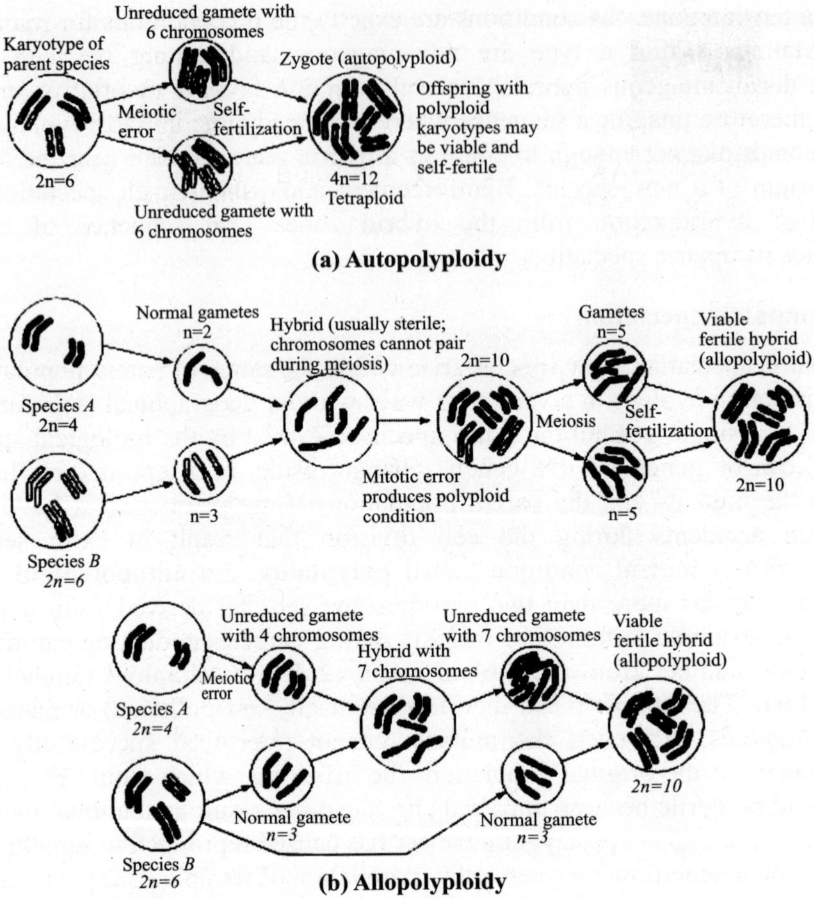

Figure 4.6. Sympatric speciation by polyploidy in plants.
(a) Autopolyploidy. (b) Allopolyploidy

Some allopolyploids are especially vigorous, perhaps because they combine the best qualities of their two parent species. Speciation of polyploids, especially allopolyploids, accounts for 25% to 50% of plant species. Some of these new polyploidy species have originated and spread within historic times. An example is a new species of salt-marsh grass that originated along the coast of southern England in the 1870s. It is an allopolyploid derived from European species (*Spartina maritima*) and an American species (*Spartina alternaflora*). Seeds of the American species, stowaways in the ballast of ships, were accidently introduced to England in the early nineteenth century. The invaders hybridized with the local species, and eventually a third species (*Spartina anglica*), morphologically distinct and reproductively isolated from its two parent species, evolved as an allopolyploid. Chromosome numbers are consistent with this mechanism of speciation; For *S. maritima*, $2n = 60$; for *S. alternaflora*, $2n = 62$; and for the new species, *S. anglica*, $2n = 122$. Since its inception, the new marsh grass has spread around the coast of Great Britain, clogging estuaries and becoming something of a pest.

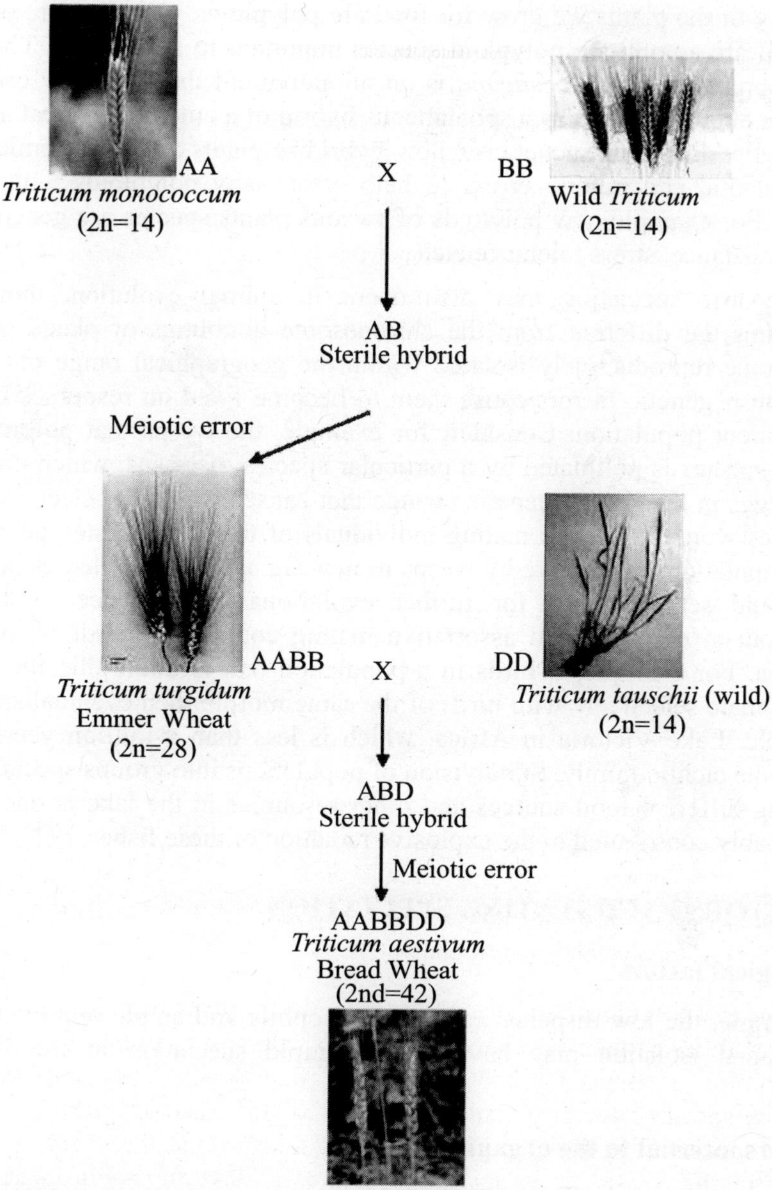

Figure 4.7. Evolution of Bread Wheat (*Triticum aestivum*). Bread wheat is the evolutionary product of two hybridization/meiotic error episodes producing allopolyploids. The first sympatric speciation produced emmer wheat, derived from hybridization between wild Triticum and a domesticated species (*Triticum monococcum*) that has been cultivated for food in the Middle East for at least 11,000 years. A second hybridization, between emmer wheat and a wild species (*T. tauschii*), gave rise to bread wheat about 8000 years ago. The letters A, B, and D represents chromosomes that can be traced to a particular species. Thus, bread wheat has chromosomes derived from three different ancestral species.

Many of the plants we grow for food are polyploids. Oats, cotton, potatoes, and wheat are among the polyploid species important to agriculture. The wheat used for bread, *Triticum aestivum*, is an allopolyploid that probably originated more than 8000 years ago as a spontaneous hybrid of a cultivated wheat and wild grass (Figure 4.7). Plant geneticists now hybridize plants and use chemicals that induce meiotic and mitotic errors to help create new polyploids with desired qualities. For example new polyloids of various plants species are generated for disease resistance, stress tolerance etc.

Sympatric speciation may also occur in animal evolution, though the mechanisms are different from the chromosome doublings of plants. Animals may become reproductively isolated within the geographical range of a parent population if genetic factors cause them to become fixed on resources not used by the parent population. Consider, for example, the wasps that pollinate figs. Each fig species is pollinated by a particular species of wasps, which mates and lays its eggs in the figs. A genetic change that caused wasps to select a different fig species would segregate mating individuals of this new phenotype from the parent population. This move by wasps to new fig species is called as *host shift.* This would set the stage for further evolutionary divergence. A balanced polymorphism coupled with assortative mating could also result in sympatric speciation. For example, if birds in a population that is dimorphic for bill size began to mate selectively with birds of the same morph, then eventual speciation is possible. Lake Victoria in Africa, which is less than a million years old, is home to the cichlid family. Subdivision of populations into groups specialized for exploiting different food sources and other resources in the lake is one process that probably contributed to the explosive radiation of these fishes.

4.4 FACTORS FACILITATING SPECIATION

1. Ecological factors

For example, the low dispersal rates of Drosophila and ample opportunities for geographical isolation may have enabled rapid speciation in the Hawaiian islands.

2. Factors internal to the organism

These are in general poorly understood, but one instructive example exists. Primitive frogs have a short inner ear that restricts them to a small range of frequencies (and thus to a limited potential to distinguish mating calls). "Advanced" frogs have a much longer inner ear and frequency range. Speciation rates seem to have increased in frog evolution, which may be facilitated by increased opportunities for sexual selection, sexual isolation, and speciation.

3. Adaptive divergence

Two populations which adapt to different environments accumulate differences in the frequencies of alleles and genotypes. During this gradual adaptive divergence of the two gene pools, reproductive barriers may evolve between the two populations. Evolution of reproductive barriers would differentiate these populations into two species. A key point in evolution by divergence is that reproductive barriers can arise without being favored directly by natural selection. Divergence of two populations is due to their adaptation to separate environments, with reproductive isolation being a secondary development. Gradual genetic divergence of two populations may also result in the evolution of pre-zygotic barriers. For instance, an ecological barrier to inbreeding may secondarily result from the adaptation of an insect population to a new host plant different from the original population's host. Here we shall look at the genetic reasons. It is probably that the characters influencing prezygotic isolation. The correlation could exist for two reasons: pleiotropy and hitch-hiking.

Pleiotropy means that one gene influences more than one phenotypic character of the organism. Consider, for example, a gene that influences the shape of a bird's beak. Beak size is related to the food that the bird eats: smaller beaks are adapted to eat smaller seeds; larger beaks are adapted to eat large seeds. If two populations occupied two islands with different-sized seeds, the populations would evolve apart as the birds adapted to the local food supply. Beak shape, in this sense, is an ecological adaptation.

Beak shape can also influence reproductive behaviour. Some birds may choose their mates by direct physical inspection of their beaks, but the influence may often be less direct. Beak shape is associated with the kind of the bird sings. Species with large beaks, for instance, do not produce rapid trills, whereas species with small beaks do. This may be a direct physical consequence of the beak size as it may be physically harder for a bird with a large beak to sing rapid trill than for a bird with a small beak. Then, when two populations adapt to different food supplies, their songs will change too. Darwin's finches partly choose their mates according to the songs they sing. Thus a change in diet can incidentally cause a change in reproductive isolation. The genetic mechanism is pleiotropy: a gene is favoured because it improves ecological adaptation will also cause some reproductive isolation. The pleiotropy arises because the same morphological character (the beak) influences both feeding and mating.

Hitch-hiking means that when natural selection favours a gene at one locus, genes at linked loci may also increase in frequency. In an experiment conducted by Dodd (1989) on fruitflies (*Drosophila pseudoobscura*), the population of the fruitfly were reared on different resources. Half of the population was placed on a starch based food medium and the remaining ones on a maltose based medium. After many generation flies had evolved detectable differences in their digestive enzymes. Dodd exploited these populations to see whether any reproductive isolation had evolved as an incidental consequence of divergence. She placed recently emerged males and females from the starch and maltose populations in a cage, after marking all the individuals of one of the populations. She then measured who mated with whom, and found that the "starch" flies preferred a "starch" mate, and the "maltose" flies a "maltose" mate. Some reproductive isolation had evolved in this case it is prezygotic isolation. It is presumably evolved because the changes that had occurred in the population influenced reproductive behaviour in some way. In Dodd's experiment, natural selection increased the frequency of genes coding for appropriate digestive enzymes. Maybe a closed linked locus influences the fruit fly's courtship dance. Then, when a gene for ecological adaptation (the digestive enzyme) increases in frequency, it may drag along with it a linked gene for a new courtship dance step. Prezygotic isolation could again evolve as a by-product, but the genetic mechanism is hitch-hiking rather than pleiotropy.

4.5 HOW MUCH GENETIC CHANGE IS REQUIRED FOR SPECIATION?

No generalizations can be made about genetic distance between closely related species. Reproductive isolation may result from changes in many loci or only in a few. Some cases involving complete reproductive isolation may result from the cumulative divergence of populations at many gene loci. In other cases, changes at only few loci produce separate species. For example, two species of Hawaiian *Drosophila*, *D. silvestris* and *D. heteroneura*, differ in allele at a gene locus determining head shape, a character important in mate recognition by these flies. However, the phenotypic effect of different alleles at this locus is amplified by at least ten other loci that interact in an epistatic system. Thus, it took more than one mutation to differentiate these two species of *Drosophila.* However, it is also clear from such examples that massive genetic change involving hundreds of loci is not mandatory for speciation. This conclusion is relevant to a debate among evolutionary biologists about the tempo of speciation.

4.5.1 Punctuated equilibrium model

In 1972, Stephen Jay Gould and Niles Eldredge published in *Models in Paleobiology* a paper whose title sounded like a war declaration: "***Punctuated Equilibria: an alternative to Phyletic Gradualism***". Eldredge and Gould distinguished two extreme hypotheses about the pattern of evolution (Figure 4.8). They named one of them *Phyletic gradualism,* which states that evolution has a fairly constant rate, that new species arise by the gradual transformation of ancestral species, and that the rate of evolution during the origin of a new species is much like that at any other time.

In reality, Gould and Eldredge had nothing of the kind in mind, and their paper was simply an attempt to show that macroevolution is a much more complex phenomenon than people had thought. They started by pointing out that species appear quickly in the fossil record and then remain unchanged for long geological periods, which is for tens of million years. The fossil record is far from perfect, but this pattern is so frequent and regular that cannot be dismissed as an exception. It is a veritable rule of the history of life, and must therefore be explained.

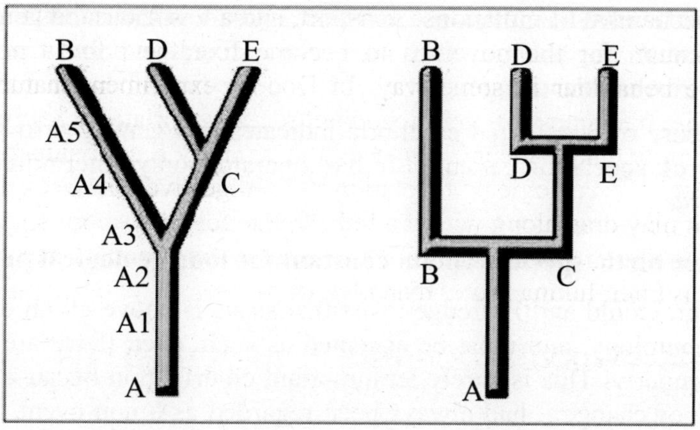

Figure 4.8. The basic patterns of species evolution according to Phyletic gradualism (left) and Punctuated equilibria. In the last model, horizontal lines indicate that speciations took place in geologically brief periods, and the fact that letters do not change along vertical lines signifies that species remained unchanged for long geological periods of time.

Gould and Eldredge looked for a model that could account for that regularity, and found it in the idea of allopatric speciation proposed by Ernst Mayr (allopatric means in another place and allopatric speciation therefore is something that species do somewhere else).

"The central concept of allopatric speciation is that new species can arise only when a small local population becomes isolated at the margin of the geographic range of its parent species. Such local populations are termed '*peripheral isolates*'. A peripheral isolate develops into a new species if '*isolating mechanisms*' evolve that will prevent the re-initiation of gene flow if the new form reencounters its ancestor at some future time. As a consequence of the allopatric theory, new fossil species do not originate in the place where their ancestors lived. It is extremely improbable that we shall be able to trace the gradual splitting of a lineage merely by following a certain species up through a local rock column".

Gould and Eldredge concluded that the history of species is not a uniform accumulation of small changes, but is made of long periods of stasis (the "equilibria") which are occasionally interrupted by quick episodes of speciation (the "punctuations"). "Evolution is like the life of a soldier: long periods of boredom and short moments of terror".

But let us examine the central points of the theory.

1. Speciation occurs rapidly in peripheral isolates

Gould and Eldredge declared that their model derives from Ernst Mayr's allopatric speciation, but this concept, in turn, was based on Sewall Wright's genetic drift. In small populations, as we have seen, genetic drift becomes the dominant mechanism of mutations' transport, and a few thousand generations are normally enough for the novelties to become fixed and for a population to undergo an irreversible change.

The theory of punctuated equilibria indicates that genetic drift was the key mechanism of speciation, even if it has operated only intermittendly in the history of life.

2. After their birth, species remain constant for long geological periods

On this point, Gould and Eldredge insist that stasis is above all an experimental fact of paleontology, and must be accepted as such, even if we are not able to explain it properly. This is surely an important contribution because stasis – i.e. the *absence* of change – had always been regarded as a non-event, and nobody before had thought that it needed explaining. Stasis, on the contrary, is an active phenomenon since species must actively work for resisting change.

The theory of punctuated equilibria, in conclusion, consists of two ideas:

(1) genetic drift had a determinant role in speciation, and

(2) stasis is a paleontological fact and must be properly explained.

At this point, however, Gould and Eldredge realized that their theory had a shattering consequence. If species actively work for their own conservation, and achieve it, we cannot say any more that they are continuously transforming themselves in order to express the features of higher taxa. This is precisely the mechanism invoked by phyletic gradualism in order to explain macroevolution, but we cannot have it both ways: stasis and phyletic gradualism are not compatible, and cannot both be true. And since it is stasis that is documented by the fossil record, Gould and Eldredge conclude that phyletic gradualism must be abandoned, even if this means that we no longer have a model for macroevolution.

The question, at this point, was *que faire*? When a widely accepted solution is found wanting, the need to replace it with an alternative model is strong, and Gould and Eldredge discovered that a way out is offered by the idea of *species selection*. If species are regarded as organic systems that are born, live and die as individuals do, one can conclude that a selection mechanism operates at the species level as it does at the individual level.

This solution takes the classical two birds with one stone. On the one hand, it is acknowledged that microevolution and macroevolution are two separate hierarchical levels, and therefore that each of them has its own independent laws. On the other hand, however, it is said that one level is ruled by the natural selection of individuals and the other by the natural selection of species, which means that all problems are solved by natural selection.

The explanation is ingenious, no doubt about that. But is it true? Unfortunately we have no proof of this, and the mechanism of macroevolution remains therefore a mystery.

4.6 POLYMOPHISM

A polymorphism is a genetic variant that appears in at least 1% of a population. Examples: the human ABO blood groups, the human Rh factor, the human major histocompatibility complex (MHC). By setting the cutoff at 1%, it excludes spontaneous mutations that may have occurred in — and spread through the descendants of — a single family.

Protein Polymorphisms: All the examples above are of the protein products of alleles. These can be identified by: serology; that is, using antibodies to detect the different versions of the protein and electrophoresis; if amino acid changes in the protein alter its net electrical charge, it will migrate more or less rapidly in an electrical field.

Enzymes are frequently polymorphic. A population may contain two or more variants of an enzyme encoded by a single locus. The variants differ slightly in their amino acid sequence and often this causes them to migrate differently under electrophoresis. By treating the gel with the substrate for the enzyme, its presence can be visualized. Electrophoretic variants of an enzyme occurring in a population are called **allozymes**.

Restriction Fragment Length Polymorphisms (RFLPs): Proteins are gene products and so polymorphic versions are simply reflections of allelic differences in the gene; that is, allelic differences in DNA. Often these changes create new — or abolish old — sites for restriction enzymes to cut the DNA. Digestion with the enzyme then produces DNA fragments of a different length. These can be detected by electrophoresis.

Single Nucleotide Polymorphisms (SNPs): Developments in DNA sequencing now make it easy to look for allelic versions of a gene by sequencing samples of the gene taken from different members of a population (or from a heterozygous individual). Alleles whose sequence reveals only a single changed nucleotide are called single nucleotide polymorphisms or SNPs. SNPs can occur in noncoding parts of the gene so they would not be seen in the protein product. might not alter the cutting site for any known restriction enzymes so they would not be seen by RFLP analysis.

Copy Number Polymorphisms (CNPs): Genetic analysis (using DNA chips and FISH) has revealed another class of human polymorphisms. These copy number polymorphisms are large (thousands of base pairs) duplications or deletions that are found in some people but not in others. On average, one person differs from another by 11 of these. One or more have been found on most chromosomes, and the list is probably incomplete. While most of this DNA is non-coding, functional genes are embedded in some of it. How, or if, the person adapts to the resulting change in gene number is unknown

How are polymorphisms useful?

Polymorphism analysis is used:

1. ***Tissue typing***; in order to find the best match between the donor, e.g., of a kidney, and the recipient.
2. Finding disease genes. Example: the gene for Huntington's disease was located when the presence of the disease was found to be linked to a RFLP whose location on the chromosome was known.
3. In population studies, for example, assessing the degree of genetic diversity in a population.
4. Determining whether two populations represent separate species or races of the same species. This is often critical to applying laws protecting endangered species.
5. Tracking migration patterns of a species (e.g., whales).

How do polymorphisms arise and persist?

1. **By mutation**. *But what keeps them in the population?* Several factors may maintain polymorphism in a population.
    a. *Founder Effect*: If a population began with a few individuals — one or more of whom carried a particular allele — that allele may come to be represented in many of the descendants. In the 1680s Ariaantje and Gerrit Jansz emigrated from Holland to South Africa, one of them bringing along an allele for the mild metabolic disease porphyria. Today more than 30000 South Africans carry this allele and, in every case examined, can trace it back to this couple — a remarkable example of the founder effect.
    b. *Genetic Drift:* An allele may increase — or decrease — in frequency simply through chance. Not every member of the population will become a parent and not every set of parents will produce the same number of offspring. The effect, called random genetic drift, is particularly strong in small populations (e.g., 100 breeding pairs or fewer); when the gene is neutral; that is, is neither helpful nor deleterious. Eventually the entire population may become homozygous for the allele or — equally likely — the allele may disappear. Before either of these fates occurs, the allele represents a polymorphism.

Two examples of reduced polymorphism because of genetic drift

By 1900 hunting of the northern elephant seal off the Pacific Coast had reduced its population to only 20 survivors. Since hunting ended, the population has rebounded from this population bottleneck to some 100,000 animals today. However, these animals are homozygous at every one of the gene loci that have been examined. Cheetahs, the fastest of the land animals, seem to have passed through a similar period of small population size with its accompanying genetic drift. Examination of 52 different loci has failed to reveal any polymorphisms; that is, these animals are homozygous at all 52 loci. The lack of genetic variability is so profound that cheetahs will accept skin grafts from each other just as identical twins (and inbred mouse strains) do. Whether a population with such little genetic diversity can continue to adapt to a changing environment remains to be seen.

4.6.1 Balanced polymorphism

In regions of the world (e.g., parts of Africa) where malaria caused by *Plasmodium falciparum* is common, the allele for sickle-cell hemoglobin is also common. This is because children who inherit one gene for the "normal" beta chain of hemoglobin and one sickle gene are more likely to survive that either homozygote. Children homozygous for the sickle allele die young from sickle-cell disease but children homozygous for the "normal" beta chain are far more susceptible to illness and death from falciparum malaria than are heterozygotes. Hence the relatively high frequency of the allele in malarial regions. When natural selection favors heterozygotes over both homozygotes, the result is **balanced polymorphism.** It accounts for the persistence of an allele even though it is deleterious when homozygous.

Another example: prion proteins: All human populations are polymorphic for the prion protein PrPC. It is encoded by the prion protein gene (PRNP). Two of the alleles have different codons at position 129: one encoding methionine; the other valine. Homozygosity for either allele increases the susceptibility to prion diseases. People who are heterozygous are more resistant. A study of elderly women who had survived the kuru epidemic of the first half of the 20th century (eating the flesh of the deceased was banned in 1950) showed that 76.7% of them were heterozygotes.

■■■■

Chapter 5

MOLECULAR EVOLUTION AND PHYLOGENY

- 5.1 Molecular Evolution
  - 5.1.1 Neutral theory of molecular evolution
  - 5.1.2 The evolution of macromolecules
  - 5.1.3 Molecular evolution is driven by changes in nucleotide
  - 5.1.4 Many mutations may be selectively neutral
  - 5.1.5 Determining and comparing the structure of macromolecules
- 5.2 Changes in Macromolecules can serve as Molecular Clocks
- 5.3 Molecular Systematics and phylogenies
  - 5.3.1 Homologous traits are inherited from a common ancestor
  - 5.3.2 Identifying ancestral traits is sometimes difficult
- 5.4 Steps in Reconstructing Phylogenies
  - 5.4.1 Morphological and developmental traits are used in reconstructing phylogenies
  - 5.4.2 Molecular traits are also useful in reconstructing phylogenies
  - 5.4.3 Reconstructing a Simple Phylogeny
  - 5.4.4 Systematists use the parsimony principle when reconstructing phylogenies
- 5.5 Biological Classification and Evolutionary Relationships
  - 5.5.1 Current biological classifications reflect evolutionary relationships
- 5.6 Phylogenetic Trees Have Many Uses
  - 5.6.1 How many times has a trait evolved?
  - 5.6.2 When did lineages split?
- 5.7 Genomes and their Evolution
  - 5.7.1 Proteins acquire new functions
  - 5.7.2 The evolution of genome size
  - 5.7.3 Gene duplication can increase genome size and complexity
- 5.8 The Uses of Molecular Genomic Information
  - 5.8.1 Molecular information is used to reconstruct phylogenies
  - 5.8.2 Molecular data are used to determine the phylogenetic histories of genes
  - 5.8.3 Molecular information provides new ways to combat diseases

5.1 MOLECULAR EVOLUTION

The laws of population genetics are usually expressed with technical terms such as mutation frequency, nucleotide-substitution rate, amino-acid-turnover rate and so on, but until the 1960s it was impossible to make direct measurements of these parameters. The only way to estimate them was by deducing their values from their visible effects on real organisms, i.e. by transferring at the molecular level what is observed in the phenotypic world.

A mutation, on the other hand, can not only be positive or negative for a given organism, but can also be neutral, in the sense that could have no adaptive value. In this case natural selection would not work on it, and its destiny would be determined by the only other existing mechanism, i.e. by genetic drift. In order to understand evolution at the molecular level, therefore, it was necessary to estimate how many neutral mutations occur in nature, on average, in respect to adaptive ones. Since direct measurements were impossible, it seemed logical to resort to indirect observations and to say that the ratio between neutral and adaptive mutations must be fairly close to the ratio between neutral and adaptive phenotypic characters.

And since neutral characters are a tiny minority, it was concluded that neutral mutations must be a very small percentage of the total. This is why Ernst Mayr concluded, in 1963, that "*it is highly unlikely that really neutral genes do exist, or that a gene could remain neutral for a long time*". Another way of obtaining indirect information on molecular evolution was offered by the study of phylogenetic trees. A typical example, in this field, is the comparison between amphibians and mammals.

Both groups derived from a common marine ancestor, but amphibians evolved much more slowly. They share so many anatomical characters that a single order comprises them all, while mammals differentiated in as many as 16 distinct orders. Mammals clearly underwent a much faster phenotypic evolution than amphibians, and it seemed logical therefore to conclude that, at the molecular level, the mutation rate has been must faster in mammals than in amphibians.

This and many other similar conclusions reinforced the idea that phenotypic evolution, molecular evolution and macroevolution took all place almost exclusively by natural selection and only in a minimal part by genetic drift. This theoretical conclusion, known as ***selectionism***, or more precisely ***panselectionism***, became increasingly popular in the 1940s and 50s, and up until the 1960s did not have any serious alternative.

5.1.1 Neutral Theory of Molecular Evolution

In the 1960s, however, it became possible to make direct measurements of the molecular parameters, and the results turned out to be quite different from those predicted by selectionism. It was found, for example, that mutations accumulate in the genomes of amphibians and mammals *at the same rate*. Phenotypic evolution did occur at very different rates in the two groups, but molecular evolution did not. It is not possible therefore to extrapolate from the phenotypic world to the molecular level, and the two levels must be kept apart. This creates of course an additional problem, because now we have to find out a bridge between the two worlds, but it is nature itself that creates the problem and we can only do our best to solve it.

In addition to mutation rate, even the other molecular parameters turned out different from the expectations of selectionism. It was discovered, for example, that neutral mutations are not in the least a tiny minority in respect to adaptive mutations, and the actual ratio is probably the other way round. At the molecular level, in other words, the dominant mechanism of evolution is not natural selection but genetic drift, and this led **Motoo Kimura** to formulate the **neutral theory of molecular evolution**.

This theory, in Kimura's own words, states,

"The great majority of evolutionary mutant substitutions are not caused by positive Darwinian selection but by random fixation of selectively neutral or nearly neutral mutants."

(It is important to underline that the adjective neutral does not mean without function. It only means that a mutation is adaptatively indifferent, i.e. it is neither better nor worse than the previous one in respect to the organism's adaptation to the environment).

Kimura's theory has not been universally accepted, and the debate between selectionism and neutralism is still going on, but the experimental data have changed for good our view of molecular evolution. Today, all biologists are aware that neutral mutations are a reality of life, and that genetic drift is, at the molecular level, at least as important, if not much more important, than natural selection. It must also be noticed that this does not diminish in the least the key role of natural selection in phenotypic evolution, and Kimura himself explicitly acknowledged that "*the basic mechanism of adaptive evolution is without doubt natural selection*".

He added however that *"underneath the remarkable procession of life and indeed deep down at the level of the genetic material, an enormous amount of evolutionary change has occurred, and is still occurring. What is remarkable, I think, is that the overwhelming majority of such change is not caused by natural selection but by random fixation of selectively neutral or nearly neutral mutants. Although such random processes are slow and insignificant for our ephemeral existence, in the span of geological times they become colossal. In this way, the footprints of time are evident in all the genomes on the earth. This adds still more to the grandeur of our view of biological evolution"*.

In 1960's, isozyme and other molecular data indicated a great deal more genetic variability in populations than anyone could have anticipated. One of the explanations for this was developed by Kimura, who suggested that much change at the molecular level is due mainly to chance (random mutation and fixation of

DNA sequences) and not selection. It indicates that some changes in genomic DNA is neutral to selection.

The theory has been cited by nonscientists as evidence of the "errors" of evolutionary theory; however, Kimura acknowledges the importance of Darwinian processes at the organism level. Kimura suggested that much molecular change takes place in the absence of selection

Two predictions

1. Rates of change of many portions of the genome will be constant
2. Highest rates of change will be observed in those portions of the genome least subject to selection

- First prediction supported by much data
    - o Proteins like cytochrome c and hemoglobin known to change at constant rates
    - o This permits timing of divergence of taxa by comparing protein structures (molecular clock)
- Second prediction supported by most data
    - o Introns (noncoding regions in DNA) more variable than exons (coding regions)
    - o Portions of genes coding for 3-D structure of proteins (which determine shape) less variable than those coding for 2-D structure
- The fact that some areas of the genome are relatively invariable indicates that selection is acting to maintain these areas as they are.
- However, the fact that most of the genomic DNA of eukaryotes is noncoding (and therefore not influenced much by selection) suggests there are many opportunities for molecular change that is neutral to selection

5.1.2 The Evolution of Macromolecules

The field of **molecular evolution** investigates the evolution of macromolecules and uses the findings to reconstruct the evolutionary history of genes and the organisms that carry them. The molecules of special interest to molecular evolutionists are nucleotides, nucleic acids, amino acids, and proteins.

Nucleic acids evolve by means of nucleotide base substitutions, which in turn can result in changes in the amino acids they encode. Alterations in the structure and function of proteins result from changes in the ordering of the amino acids of which they are composed. Molecular evolutionists characterize the precise structures of these macromolecules: nucleotide sequences in nucleic acids, and primary structure in proteins.

They use that information to determine how rapidly these macromolecules have changed and why they have changed. Phylogenetic information is essential for determining the order of changes in molecular characters because knowing the order of such changes is usually the first step in inferring their causes. Conversely, knowledge of the pattern and rate of change of a given macromolecule is crucial to attempts to reconstruct the evolutionary history of groups of organisms.

5.1.3 Molecular evolution is driven by changes in nucleotide

Molecular evolution differs from phenotypic evolution in that random genetic drift and mutation exert important influences on the rates and direction of nucleotide changes. A *mutation* is any change in the genetic material. Many mutations do not alter the protein encoded by the mutated genes. The reason is that in the "universal genetic code", most amino acids are specified by more than one codon. Leucine, for example, is specified by six different codons: UUA, UUG, CUU, CUC, CUA, and CUG. A mutation that does not change the amino acid—UUA to UUG, for example—is known as a **synonymous** or **silent mutation** (Figure 5.1a). Synonymous mutations do not affect the functioning of a protein (and hence the organism), and are, therefore, unlikely to be influenced by natural selection.

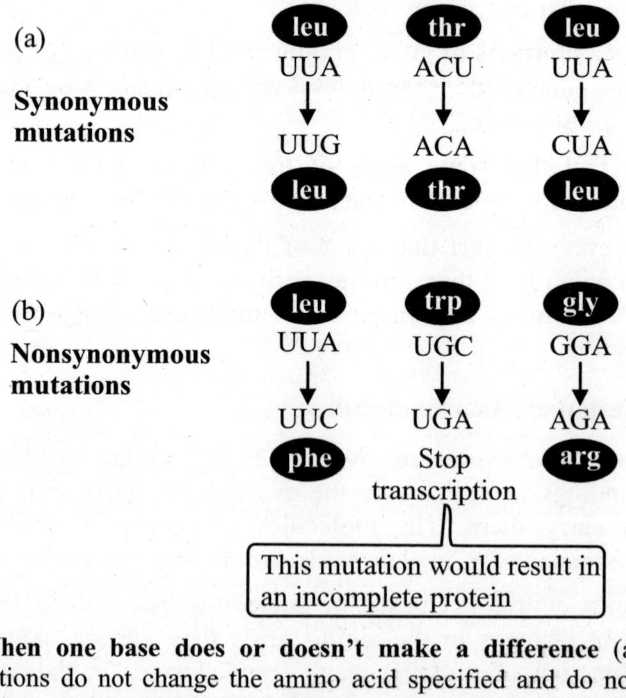

Figure 5.1. When one base does or doesn't make a difference (a) Synonymous mutations do not change the amino acid specified and do not affect protein function; such mutations are unlikely to be agents of natural selection. (b) Nonsynonymous mutations do change the amino acid sequence and are in general likely to have an effect (usually deleterious) on protein function and natural selection.

Because they are unlikely to be influenced by natural selection, synonymous mutations can spread at rates determined by rates of mutation and genetic drift, so that they completely replace one nucleotide base or longer sequence with another throughout an entire population or species. In this section we will call such replacements **substitutions** to distinguish them from a replacement of one nucleotide by another in a single individual (which is also often called a substitution). Molecular evolutionists are primarily interested in population-wide nucleotide substitutions.

A mutation that *does* change the amino acid sequence encoded by a gene is known as a **non-synonymous mutation**. For example, a change from UUA to UUC would result in phenylalanine rather than leucine in the encoded amino acid (Figure 5.1b). In general, non-synonymous mutations are likely to be deleterious to the organism. But not every amino acid change alters a protein's shape (and hence its functional properties). Therefore, some non-synonymous mutations may be selectively neutral, or nearly so.

Most natural populations of organisms harbor much more genetic variation than we would expect if genetic variation were influenced primarily by natural selection. This discovery, combined with the knowledge that many mutations do not change molecular function, stimulated the development of the neutral theory of molecular evolution.

5.1.4 Many mutations may be selectively neutral

In 1968, Motoo Kimura suggested that, at the molecular level, the majority of mutations are selectively neutral; that is, they confer neither an advantage nor a disadvantage on their bearers. Thus, the majority of evolutionary changes in macromolecules, and much of the genetic variation within species, could result neither from directional selection of advantageous alleles nor from stabilizing selection, but rather from genetic drift.

To see why this is so, consider a population of size N and a neutral mutation rate of μ per gamete per generation at a locus. The number of new mutations would, on average, be $\mu \times 2N$, because $2N$ gene copies are available to mutate in a diploid organism. According to drift theory, the probability that a mutation will be fixed by drift alone is its frequency, p, which equals $1/(2N)$ for a newly arisen (and hence very rare) mutation. Therefore, the number of neutral mutations that arise per generation that are likely to become fixed is $2N\mu \times 1/(2N) = \mu$ — which is the mutation rate.

Thus, the rate of fixation of neutral mutations is theoretically constant and is equal to the mutation rate. So if most mutations of macromolecules do not affect their functioning, macromolecules should diverge from one another at a constant rate. Indeed, in the 1960s, comparative studies of several proteins indicated that, for a particular protein, rates of amino acid substitutions were similar in all evolutionary lineages having that protein. In other words, the rate of evolution of particular proteins might be constant over time, and in effect can be a "**molecular**

clock." We will discuss molecular clocks later in this section and show how, with care, they can be used to study several features of molecular evolution.

5.1.5 Determining and Comparing the Structure of Macromolecules

To investigate patterns of molecular evolution, biologists must first determine the precise structure of molecules. The base sequences of nucleic acids provide information about the primary structure of the proteins they encode. The invention of the polymerase chain reaction technique allowed biologists to determine the sequences of regions of DNA not only from living tissues, but also from fossilized remains, mummified tissues, dried skins in museums, and pressed plants in herbaria, even though these objects contain only tiny amounts of DNA. DNA has been extracted and amplified from human fossils more than 30,000 years old and from plant leaf fossils 40,000 years old. Once the sequences of molecules from different organisms have been determined, they can be compared. The purpose of comparing them is to identify the locations of deletions, insertions, and substitutions that have occurred in the molecules of interest since the organisms diverged from a common ancestor. A simple hypothetical example illustrates how this is done. In Figure 5.2 we compare two amino acid sequences (1 and 2) from homologous proteins in different organisms. The two sequences differ in the number and identity of their amino acid residues. To compare these sequences, we first observe that, although the sequences appear quite different, they would become similar if we were to insert a gap after the first amino acid in sequence 2 (after the leucine residue). In fact, these sequences then differ only by one amino acid at position 6 (serine or phenylalanine). Insertion of a single gap—that is, correcting a deletion—*aligns* these sequences. Longer sequences and those that have diverged more extensively require more elaborate adjustments.

After we have aligned the sequences, we can compare them by counting the number of nucleotides or amino acids that differ between them. If we add more sequences to our original example and sum the number of similar and different amino acids in each pair of sequences, we can construct a *similarity matrix* (Figure 5.2), which gives us a measure of the changes that have occurred during the divergence of the organisms.

The longer the molecules have been evolving separately, the more differences they should have. Enough analyses of mammalian genes have been performed to show that the rate of nonsynonymous nucleotide substitution varies from nearly zero to about 3×10^{-9} substitutions per locus per year. Synonymous substitutions in the protein-coding regions of genes have occurred about five times more rapidly than nonsynonymous substitutions. In other words, substitution rates are highest at nucleotide positions that *do not change the amino acid being expressed* (Figure 5.3). The rate of substitution is even higher in **pseudogenes**, duplicate copes of genes that have undergone one or more mutations that eliminate their ability to be expressed. Why are these rates of substitution so dissimilar?

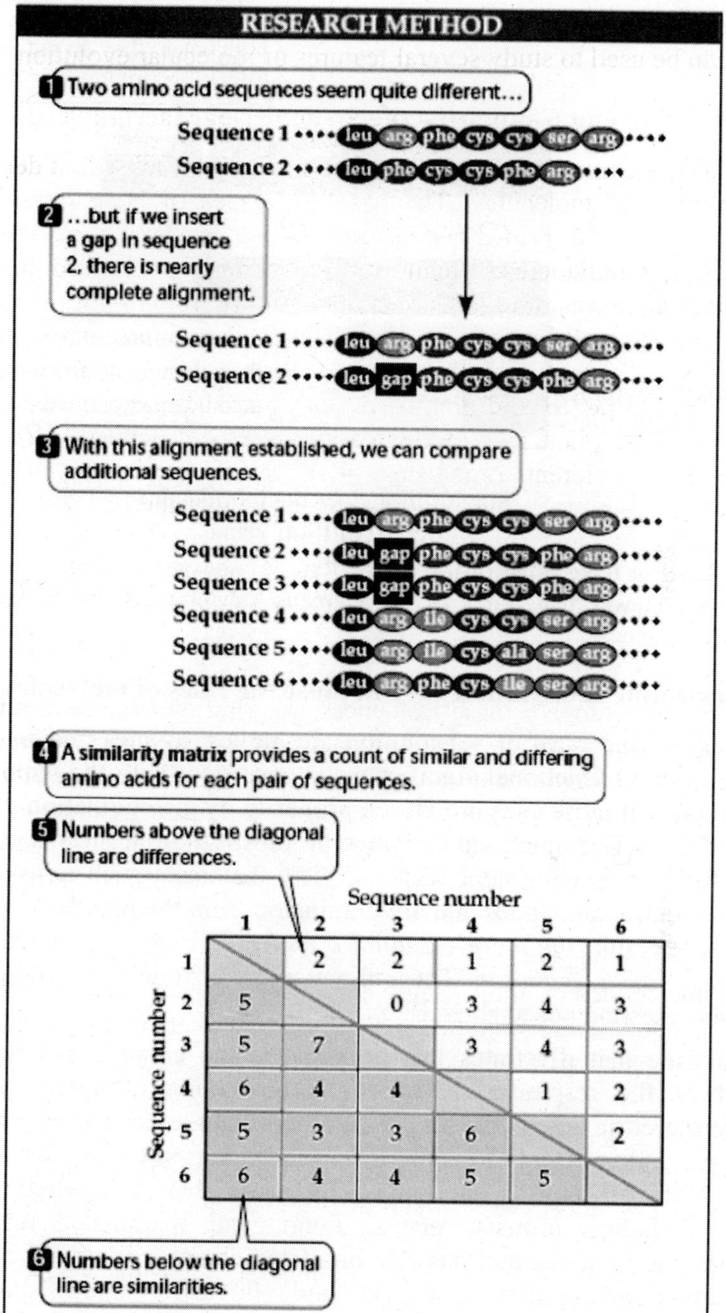

Figure 5.2. Amino acid sequence alignment. Insertion of a gap allows us to align two homologous amino acid sequences so that we can compare them. Once the alignment is established, sequences from more organisms can be added and compared. A similarity matrix sums similarities and differences between each pair of organisms. The larger the number of similarities, the more recent the presumed common ancestor of the organisms.

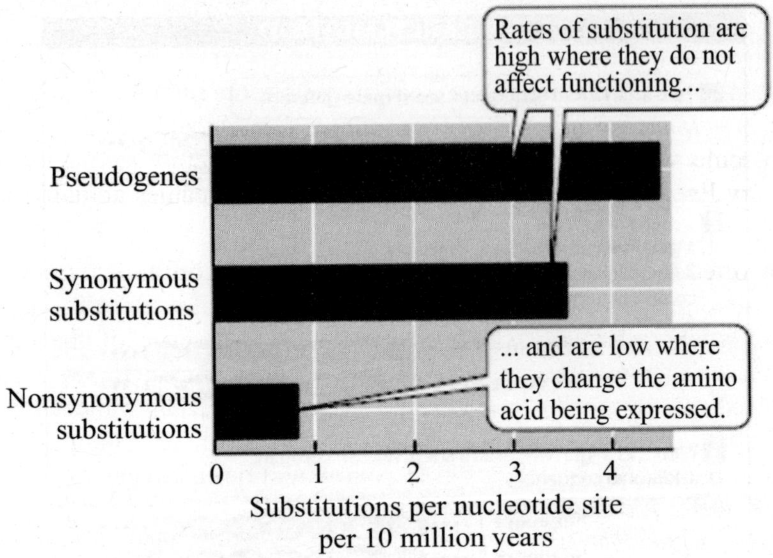

Figure 5.3. Rates of Base substitution differ. Rates of nonsynonymous substitutions are much slower than rates of synonymous substitutions and substitutions in pseudogenes.

Rates of nucleotide substitution vary because the roles of molecules differ

The observation that rates of substitution are highest at sites and in molecules where they have no functional significance is consistent with the hypothesis that substitution rates at these sites are driven primarily by a combination of mutation and genetic drift. The much slower rates of substitution at sites that *do* affect molecular function is consistent with the view that most such nonsynonymous mutations are disadvantageous and are eliminated from the population by natural selection. As a result, *the more essential a molecule is for cell functioning, the slower the rate of its evolution*. These functional constraints provide part of the answer to the question posed above.

A molecule that illustrates this principle is the enzyme cytochrome *c*, a component of the respiratory chain of mitochondria. Together with other enzymes of the citric acid cycle and respiratory chain, cytochrome *c* is found in all eukaryotes and is essential for the life of the eukaryotic cell. The amino acid sequences of cytochrome *c* are known for more than a hundred species of organisms, including protists, plants, fungi, and mammals. Within these cytochromes *c* are regions that have accumulated changes relatively quickly; for example, amino acid positions 44, 89, and 100 differ among many of the organisms compared. There are also invariant amino acid positions, such as 14, 17, 18, and 80. This particular set of invariant residues is known to interact with the iron-containing heme group, which is essential for enzyme functioning. Because any mutations that changed these amino acids would have diminished the functioning of cytochrome *c*, they would have been removed by natural selection when they arose.

5.2 CHANGES IN MACROMOLECULES CAN SERVE AS MOLECULAR CLOCKS

Earlier we mentioned that the rate of evolution of some macromolecules might in effect be a *molecular clock*. To function as a molecular clock, a particular macromolecule would need to evolve at an approximately constant rate in all evolutionary lineages that possess it. But do macromolecules actually behave in this way?

They often do. For example, if we use the fossil record to determine the time since the divergence of certain organisms, and then plot this time against the number of amino acids by which the nucleotide sequence of the organisms' cytochrome *c* proteins differ, we find that differences in cytochrome *c* sequences have evolved at a relatively constant rate (Figure 5.4). Many other proteins also show constancy in the rate at which they have accumulated changes over time. Analyses of patterns of molecular evolution would be relatively easy if the rates of change were the same for all macromolecules. Unfortunately, molecular clocks tick at somewhat different rates over evolutionary time. By comparing the rates of a variety of molecular clocks, insights can be gained into why different molecules have evolved at such different rates. Such differences exist because proteins differ in the nature of the functional constraints on their evolution.

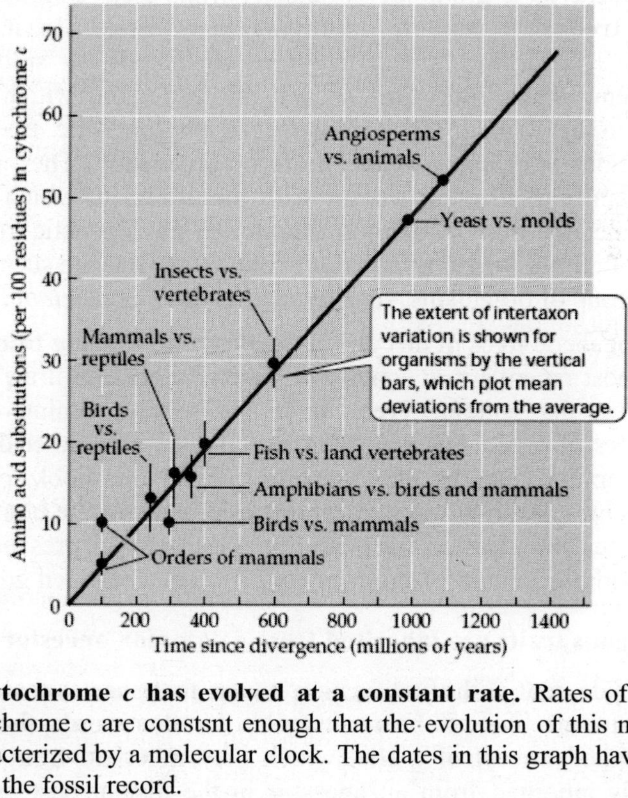

Figure 5.4. Cytochrome *c* has evolved at a constant rate. Rates of substitution in cytochrome c are constsnt enough that the evolution of this molecule can be characterized by a molecular clock. The dates in this graph have been inferred from the fossil record.

For example, the rate of evolution of an enzyme would change drastically if a mutation meant that the enzyme lost its function, or if the population of the species in which the enzyme was found changed dramatically in size.

In addition, rates of molecular evolution are faster in organisms with short generation times than in organisms with long generation times. Short-generation organisms have more rounds of DNA replication, and thus more opportunities for errors in replication, per unit of time than long-generation organisms. For example, the rate of substitution per base per year in introns is from two to four times greater in rodents (which may reach reproductive age within 6 months and have several generations per year) than in primates (which have multi-year pre-reproductive periods and multiyear generation times).

5.3 MOLECULAR SYSTEMATICS AND PHYLOGENIES

Molecular systematics is the use of molecular data as characters to assess phylogenetic position of taxa for classification. It generally uses sequence data from specific genes (rDNA, mtDNA, cpDNA, etc.), in comparison to sequence across taxa which reveals patterns of divergence in the past. Molecular clocks are inferred to establish timing of divergence when possible.

A **phylogeny** is a hypothesis proposed by a systematist that describes the history of descent of a group of organisms from their common ancestor. A **phylogenetic tree** is a way of portraying that history. In it, a lineage is represented as a branching "tree", in which each node, or split, represents a speciation event. Thus, the tree shows the order in which lineages are hypothesized to have split. To reconstruct a phylogenetic tree, systematists analyze evolutionary changes in the traits of organisms. They are guided by Darwin's fundamental idea of descent with modification, which states that all species are descended from a common ancestor. A phylogenetic tree may portray the evolution of all life forms; of a major evolutionary lineage such as the insects; or of a small group of organisms, such as snail genus *Oncomelania* (Figure 5.5).

The phylogenetic trees in this chapter depict time flowing from left (earliest) to the right (most recent); it is equally common practice to draw trees with the earliest times at the bottom. Trees show the relative timing of separations between lineages of organisms, but, unless combined with other data, we do not date those separations. In the phylogenetic trees in this book, position on the horizontal axis has meaning, but vertical distance between the branches does not. Vertical distances are adjusted for legibility and clarity of presentation; they do not correlate with the degree of similarity or difference between groups.

5.3.1 Homologous traits are inherited from a common ancestor

The process of descent with modification means that species that share a recent common ancestor are likely to be very similar. In other words, they should share many traits, called **ancestral traits**, which they inherited from their common ancestor. Traits inherited from an ancestor in the distant past are likely to be shared by a large number of species. Traits that first appeared in a more recent ancestor should be shared by fewer species. But in all cases, the sharing of traits by a group of species indicates that they are likely to be descendants of a common ancestor.

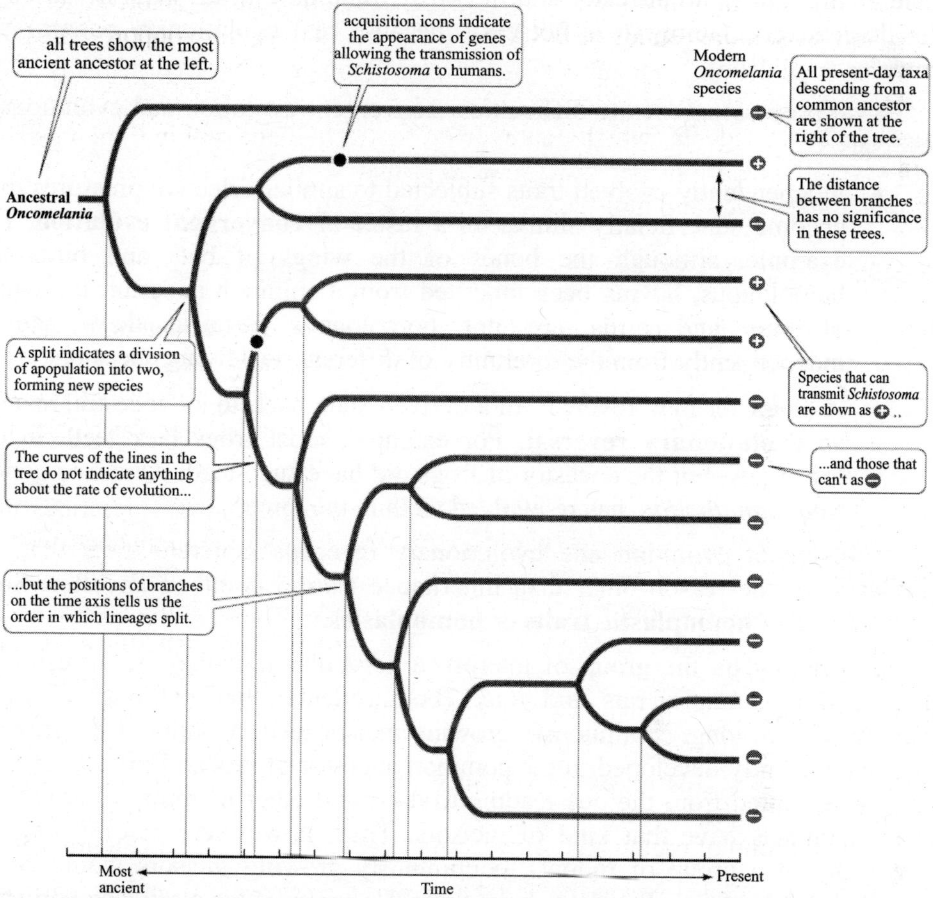

Figure 5.5. How to read a Phylogenetic tree. A phylogenetic tree displays the order in which lineage split. This example shows the phylogeny of *Oncomelania* snails, the intermediate hosts of the human parasite *Schistosoma*. Acquisition of the ability to transmit *Schistosoma* occurred twice during the evolution of this lineage.

Any features shared by two or more species that have been inherited from a common ancestor are said to be **homologous**. These features may be any heritable traits, including anatomical structures, behavior patterns, and DNA sequences. Traits that are shared by most or all of the organisms in a lineage of interest are likely to have been inherited relatively unchanged from an ancestor that lived very long ago. For example, all living vertebrates have a vertebral column, all known fossil vertebrates had a vertebral column, and all vertebrates are descended from the same common ancestor. Therefore, the vertebral column is judged to be homologous in all vertebrates.

A trait that differs from its ancestral form is called a **derived trait**. To determine how traits have changed during evolution, systematists must infer the state of the trait in an ancestor and then determine how it has been modified in the descendants. Doing so is not easy, because real evolutionary patterns are complex.

Two processes generate difficulties: convergent evolution and evolutionary reversals.

- Independently evolved traits subjected to similar selective pressures may become superficially similar as a result of **convergent evolution**. For example, although the bones of the wings of bats and birds are homologous, having been inherited from a common ancestor, the wings of bats and birds are not homologous because they evolved independently from the forelimbs of different non-flying ancestors.
- A character may revert from a derived state back to an ancestral state—an **evolutionary reversal**. For example, most frogs lack teeth in the lower jaw, but the ancestor of frogs did have such teeth. One frog genus, *Amphignathodon*, has re-evolved teeth in the lower jaw.

Convergent evolution and evolutionary reversals generate traits that are similar for some reason other than inheritance from a common ancestor. Such traits are called **homoplastic traits** or **homoplasies**.

Depending on the group of interest, a particular trait may be ancestral or derived. For example, rats and mice (both rodents), but not dogs or other mammals, have long, continuously growing incisor teeth. Continuously growing incisors evidently developed in the common ancestor of rats and mice after their lineage separated from the one leading to dogs and other mammals, because no other mammals have that kind of incisors. Thus, if we were reconstructing a phylogeny of a group of rodents, continuously growing incisors would be an ancestral trait because all rodents have them. However, if we were reconstructing a phylogeny of all mammals, continuously growing incisors would be a derived trait unique to the rodents. The distinction between ancestral and derived traits is important in reconstructing phylogenies.

5.3.2 Identifying ancestral traits is sometimes difficult

Distinguishing derived traits from ancestral traits may be difficult because traits often become so dissimilar that ancestral states are unrecognizable. The leaves of plants, for example, have diverged to form many different structures. Several lines of evidence, especially details of their structure and development, indicate that protective spines, tendrils, and brightly colored structures that attract pollinators are all modified leaves; these structures are *homologs* of one another even though they do not resemble one another closely.

One method of distinguishing ancestral traits from derived traits is to assume that an ancestral trait should be found not only among the species of the **ingroup** (the lineage of interest), but also in outgroups. An **outgroup** is a lineage that is closely related to the ingroup, but which branches off from the ingroup before its base on the evolutionary tree. Traits found only within the ingroup, on the other hand, are likely to be derived traits.

5.4 STEPS IN RECONSTRUCTING PHYLOGENIES

The first step in reconstructing a phylogeny is to select the group of organisms whose phylogeny is to be reconstructed— the ingroup—and an appropriate outgroup. The next step is to choose the characters that will be used in the analysis and to identify the possible forms of those characters (states or traits). A trait may be the presence or absence of a character, or one of the states a particular character may have, such as the number of body segments or number of appendages. The next, and usually the most difficult step, is to determine which traits are ancestral and which are derived. Finally, through phylogenetic analysis, systematists must distinguish homologous from homoplastic traits. Because organisms differ in many ways, systematists use many characters to reconstruct phylogenies. Some of these characters, such as morphology, are readily preserved in fossils; others, such as behavior and molecular structures, rarely survive fossilization processes. Systematists use physiological, behavioral, molecular, and structural characters that can be assessed in both living and fossil organisms. The more characters that are measured, the more likely it is that the phylogeny will reflect the actual evolutionary pattern.

5.4.1 Morphological and developmental traits are used in reconstructing phylogenies

An important source of information for systematists is **morphology**— that is, the sizes and shapes of body parts. Since living organisms have been studied for centuries, we have a wealth of recorded morphological data, as well as extensive museum and herbarium collections of organisms whose traits can be measured. Technological tools, such as the electron microscope and computer simulations, enable systematists to measure and analyze the structures of organisms at much finer scales, down to the level of molecules, than was formerly possible. The early developmental stages of many organisms reveal similarities to other organisms, but those similarities may be lost during later development. For example, the larvae of marine creatures called sea squirts have a rod in the back—the *notochord*—that disappears as they develop into adults. All vertebrate animals also have a notochord at some time during their development (Figure 5.6). This shared structure is one of the reasons for believing that sea squirts are more closely related to vertebrates than would be suspected if only adult sea squirts were examined.

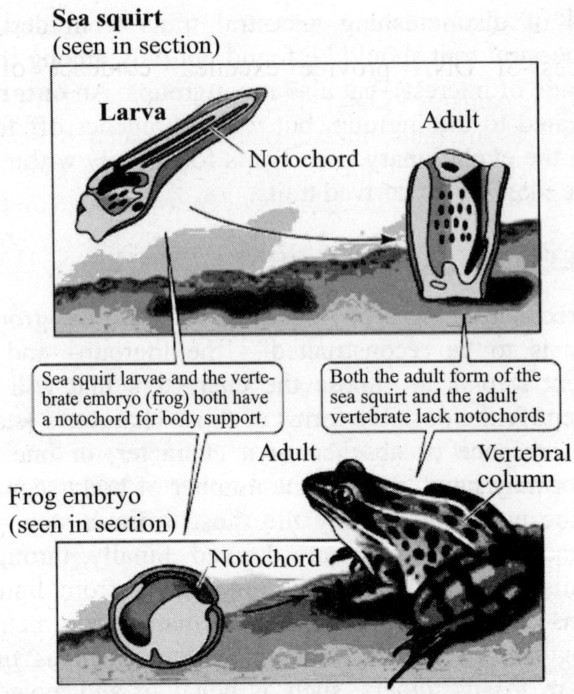

Figure 5.6. A larva reveals evolutionary relationships. Sea squirt larvae, but not adults, have a well developed notochord that reveals their evolutionary relationship to vertebrates, all of which have a notochord at some time during their life cycle. In adult vertebrates, the vertebral column replaces the notochord as the support structure.

Fossils show us where and when organisms lived in the past and give us an idea of what they looked like. Fossils provide important evidence that helps us distinguish ancestral from derived traits. The fossil record also reveals when lineages diverged and began their independent evolutionary histories. However, few or no fossils have been found for some groups whose phylogeny we may wish to determine.

5.4.2 Molecular traits are also useful in reconstructing phylogenies

The molecules that make up organisms are also heritable traits that may diverge among lineages over evolutionary time. Molecular traits that are most useful for constructing phylogenies: the primary structures of proteins and nucleic acids (DNA and RNA).

Protein primary structure

Relatively precise information about phylogenies can be obtained by comparison of the primary structures of proteins. We can measure genetic differences between two lineages by obtaining homologous proteins from both of them and determining the number of amino acids that have changed since the lineages diverged from a common ancestor.

DNA base sequence

The base sequences of DNA provide excellent evidence of evolutionary relationships among organisms. The cells of eukaryotes have genes in their mitochondria as well as in their nuclei; plant cells also have genes in their chloroplasts. The chloroplast genome (cpDNA), which is used extensively in phylogenetic studies of plants, has changed little over evolutionarily time. Mitochondrial DNA (mtDNA) has been used extensively for studies of evolutionary relationships among animals.

Relationships among the apes were investigated by sequencing more than 10,000 base pairs making up a segment of DNA that includes a hemoglobin pseudogene (a nonfunctional DNA sequence derived early in primate evolution by duplication of a hemoglobin gene). The outgroups for the analysis were *Ateles*, the spider monkeys of tropical America, and *Macaca*, the Rhesus monkey of Asia. The DNA data strongly indicate that chimpanzees and humans share a more recent ancestor than they do with gorillas, a conclusion supported by other types of molecular data.

5.4.3 Reconstructing a Simple Phylogeny

To show how a phylogeny is constructed, let's consider eight vertebrate animals: the lamprey, perch, pigeon, chimpanzee, salamander, lizard, mouse, and crocodile. We will assume initially that a given derived trait evolved only once during the evolution of these animals, and that no derived traits were lost from any of the descendant groups (Table 5.1). For simplicity, we have selected traits that are either present (+) or absent (–). A group of jawless fishes called the lampreys is thought to have separated from the lineage leading to the other vertebrates before the jaw arose. Therefore, we will choose the lamprey as the outgroup for our analysis. Derived traits are those that have been acquired by other members of the lineage since they separated from the lamprey.

We begin by noting that the chimpanzee and mouse share two unique traits: mammary glands and fur. Those traits are absent in both the outgroup and the other species of the ingroup. Therefore, we infer that mammary glands and fur are derived traits that evolved in a common ancestor of chimpanzees and mice after that lineage separated from the ones leading to the other vertebrates. In other words, we provisionally assume that mammary glands and fur evolved only once among the animals in our ingroup.

The pigeon has one unique trait: feathers. As before, we provisionally assume that feathers evolved only once, after the lineage leading to birds separated from that leading to the mouse, chimpanzee, and crocodile. By the same reasoning, we assume that keratinous scales evolved only once, after the lineage leading to crocodiles, birds, and lizards separated from the lineage leading to mammals. We assume that claws or nails evolved only once, after the lineage leading to salamanders separated from the lineage leading to those animals that have claws or nails. We make the same assumption for lungs and jaws, continuing to minimize the number of evolutionary events needed to produce the patterns of shared traits among these eight animals.

Table 5.1. Table showing shared derived traits in eight vertebrate orders.

1 Eight Vertebrates Ordered According to Unique Shared Derived Traits								
	Derived trait*							
Taxon	**Jaws**	**Lungs**	**Claws or Nails**	**Gizzard**	**Feathers**	**Fur**	**Mammary glands**	**Keratinous scales**
Lamprey (outgroup)	–	–	–	–	–	–	–	–
Perch	+	–	–	–	–	–	–	–
Salamander	+		–	–	–	–	–	–
Lizard	+	+	+	–	–	–	–	+
Crocodile	+	+	+	+	–	–	–	+
Pigeon	+	+	+	+	+	–	–	+
Mouse	+	+	+	–	–	+	+	–
Chimpanzee	+	+	+	–	–	+	+	–

*A plus sign indicates the trait is present, a minus sign that it is absent

Using this information, we can reconstruct a provisional phylogeny (Figure: 5.7). We assume that the animals that share unique derived traits have a common ancestor not shared with the animals lacking those traits. We assume, for example, that mice and chimpanzees, the only two animals that share fur and mammary glands, share a more recent common ancestor with each other than they do with birds and crocodiles. Otherwise, we would need to assume that the ancestors of birds and crocodiles also had fur and mammary glands, but that they subsequently lost them—unnecessary additional assumptions.

This phylogeny was constructed from the information given in Table 5.1 using the parsimony principle. Except for the lamprey. All of the animals display at least one derived trait.

Figure 5.7 shows a phylogeny for these eight vertebrates, based on the traits we used and the assumption that each derived trait evolved only once. This particular phylogeny was easy to construct because the animals and traits fulfilled the assumptions that derived traits appeared only once in the lineage and that they were never lost after they appeared. If we had included a snake in the group, our second assumption would have been violated, because the lizard ancestors of snakes had limbs that were subsequently lost (along with their claws). We would need to examine additional traits to determine that the lineage leading to snakes separated from the one leading to lizards long after the lineage leading to lizards separated from the others. In fact, the analysis of a number of traits shows that snakes evolved from burrowing lizards that became adapted to a subterranean existence.

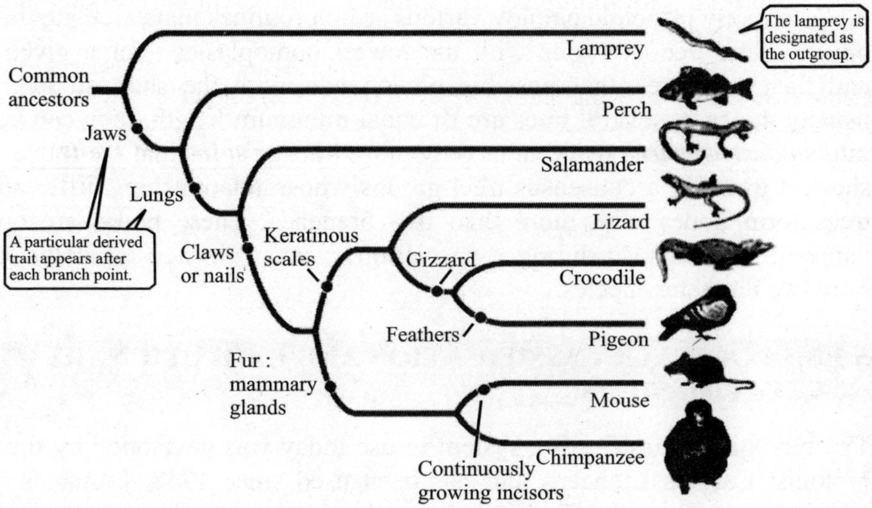

Figure 5.7. A probable phylogeny of eight vertebrates.

5.4.4 Systematists use the parsimony principle when reconstructing phylogenies

The simple method we used to reconstruct our vertebrate phylogeny does not work in the vast majority of cases because we know from fossil and other evidence that traits can change more than once or undergo evolutionary reversal. Several methods are used to deal with these complexities. The most widely used ones employ the **parsimony principle**. In its most general form, the parsimony principle states that one should prefer the simplest hypothesis that is capable of explaining the observed data. Its application to the reconstruction of phylogenies means minimizing the number of evolutionary changes that need to be assumed over all characters in all groups in the tree. In other words, the best hypothesis is one that requires the fewest homoplasies.

Using the parsimony principle is appropriate, not because all evolutionary changes occurred parsimoniously, but because it is generally wiser to adopt the simplest explanation that can account for the observed data. More complicated explanations are accepted only when the evidence requires them. As we mentioned earlier, phylogenetic trees are hypotheses about evolutionary relationships. They are continually modified as additional traits are measured and as new fossil evidence becomes available.

Another method of reconstructing phylogenies, called the *maximum likelihood method*, is used primarily with molecular data. The computer programs employed in this method are complex. They are designed to deal with the fact that mutations commonly change nucleotide sequences.

Whatever method is employed, reconstructing the phylogeny for any group of organisms is difficult. For example, there are 34,459,425 possible phylogenetic trees for a lineage with only 11 species! Computer programs using the parsimony principle employ various search routines that calculate the shortest phylogenetic tree—the one with the fewest homoplasies—for a given data set and then compare other possible phylogenies with the shortest one. If, as is usually the case, several trees are of equal minimum length, they can be merged into a *consensus tree* that retains only those lineage splits that are found in all the shortest trees. In a consensus tree, groups whose relationships differ among the trees form nodes with more than two branches. These nodes are considered "unresolved" because during a speciation event, a lineage typically splits into only two daughter species.

5.5 BIOLOGICAL CLASSIFICATION AND EVOLUTIONARY RELATIONSHIPS

The biological classification system in use today was developed by the Swedish biologist Carolus Linnaeus and has been used since 1758. Linnaeus's system, referred to as *binomial nomenclature*, allows scientists throughout the world to refer unambiguously to the same organisms by the same names.

Linnaeus gave each species two names, one identifying the species itself and the other the genus to which it belongs. A **genus** (plural, genera; adjectival form, generic) is a group of closely related species. In many cases, the name of the taxonomist who first proposed the species name is added at the end. Thus, *Homo sapiens* Linnaeus is the name of the modern human species. *Homo* is the genus, to which the species belongs, and *sapiens* identifies the particular species in the genus *Homo*; Linnaeus proposed the species name *Homo sapiens*. You can think of the generic name *Homo* as equivalent to your surname and the specific name *sapiens* as equivalent to your first name. The generic name is always capitalized; the name identifying the species always lowercased. Both names are always italicized, whereas common names of organisms are not. Rather than repeating a generic name when it is used several times in the same discussion, biologists often spell it out only once and abbreviate it to the initial letter thereafter (for example, *D. melanogaster* is the abbreviated form of *Drosophila melanogaster*). Recognizing and interpreting similarities and differences among organisms is easier if the organisms are classified into groups that are ordered and ranked.

Any group of organisms that is treated as a unit in a biological classification system, such as the genus *Oncomelania*, or all snails, is called a **taxon** (plural, taxa). In the Linnaean system, species and genera are further grouped into a hierarchical system of higher taxonomic categories. The taxon above the genus in the Linnaean system is the **family**. The names of animal families end in the suffix "-idae." Thus, Formicidae is the family that contains all ant species, and the family Hominidae contains humans and our recent fossil relatives, as well as chimpanzees and gorillas. Family names are based on the name of a member genus; Formicidae is based on the genus *Formica*, and Hominidae is based on *Homo*. Plant classification follows the same procedures, except that the suffix "-aceae" is used with family names instead of "-idae."

Thus, Rosaceae is the family that includes the genus of roses (*Rosa*) and its close relatives. Families, in turn, are grouped into **orders**, orders into **classes**, and classes into **phyla** (singular, phylum). The phyla of plants, fungi, and animals are grouped into the **kingdoms** Plantae, Fungi, and Animalia.

Biological classification systems and the unique names they provide for organisms are important for several reasons. They improve our ability to infer relationships among organisms. They are also an aid to memory and precise communication. It is impossible to remember the characteristics of many different organisms unless we can group them into categories based on shared characteristics. Classification systems are also useful as predictors. For example, the discovery of biochemical precursors of the drug cortisone in certain yams of the genus *Dioscorea*, stimulated a successful search for higher concentrations of the drug in other *Dioscorea* species.

5.5.1 Current biological classifications reflect evolutionary relationships

Biological classification systems are designed to express relationships among organisms. The kind of relationship we wish to express influences which features we use to classify organisms. If, for instance, we were interested in a system that would help us decide what plants and animals were desirable as food, we might devise a classification based on tastiness, ease of capture, and the type of edible parts each organism possessed. Early classifications of plants were designed according to these criteria. Biologists do not use such systems today, but those systems served the needs of the people who developed them.

Most taxonomists today believe that biological classification systems should reflect the evolutionary relationships of organisms, and that taxonomic groups should be **monophyletic**.

A monophyletic group (also called a **clade**) contains all the descendants of a particular ancestor and no other organisms.

A group containing some members that do not share the same common ancestor is said to be **polyphyletic**.

A group that contains some, but not all, of the descendants of a particular ancestor is said to be **paraphyletic** (Figure 5.8).

A monophyletic group can be removed from a phylogenetic tree by a single "cut" in the tree, as shown in Figure 5.8. Taxonomists agree that polyphyletic groups are inappropriate as taxonomic units. The classifications used today still contain many polyphyletic groups because many organisms have not been studied well enough to distinguish between characters that are homologies and those that are homoplasies. However, as soon as they detect such homoplasies, taxonomists change their classifications to eliminate polyphyletic taxa.

Although most taxonomists prefer strict phylogenetic classifications, some believe that classification systems should also reflect degrees of difference among organisms. According to this view, certain paraphyletic groups that have undergone rapid evolutionary change and diversification and evolved distinctive characters should be retained. Such groups are called **Grades**. This perspective can be illustrated using birds, crocodiles, and their relatives.

We have some molecular evidence that birds and crocodilians (a group that includes crocodiles and alligators) share a more recent common ancestor than crocodilians and turtles share with snakes and lizards (Figure 5.9*a*). Traditionally, crocodilians are grouped with snakes, lizards, and turtles in the class Reptilia. Birds are placed in a separate class, Aves (Figure 5.9*b*). This classification came about because crocodilians have evolved more slowly than birds since the two lineages separated. Birds rapidly evolved significant and distinctive adaptations for flight, but crocodilian traits have changed little from those of their ancestors. As a result, crocodilians are more similar in many ancestral features to snakes and lizards than they are to birds. They look like, and are physiologically similar to, very large lizards.

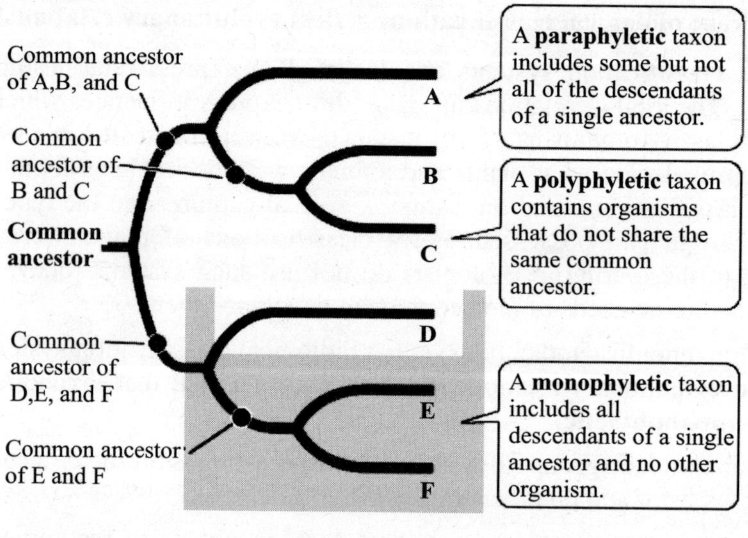

Figure 5.8. Monophyletic, Polyphyletic, and Paraphyletic taxa. Monophyletic groups are preferred by most taxonomists. Polyphyletic groups are considered inappropriate as taxonomic units, but taxonomists sometimes do use paraphyletic taxa.

Thus, the traditional class Reptilia is paraphyletic because it does not include all the descendants of its common ancestor; that is, birds are excluded (Figure 5.9*b*). If only monophyletic taxa were permitted, birds would be grouped with crocodilians in a single taxon separate from snakes and lizards or birds would be included within an expanded class Reptilia. Retaining birds as a separate class (that is, retaining reptiles as a paraphyletic group) emphasizes that birds have evolved unique derived traits since they separated from reptiles, and thus are a distinct grade.

Although the current preference is to change classifications to eliminate paraphyletic groups, some of the most familiar taxonomic categories—

gymnosperms and reptiles, for example — are paraphyletic. Because of their familiarity and the extensive literature devoted to them, these categories are likely to remain in use even after their formal taxonomic designations have been changed to better represent their evolutionary relationships.

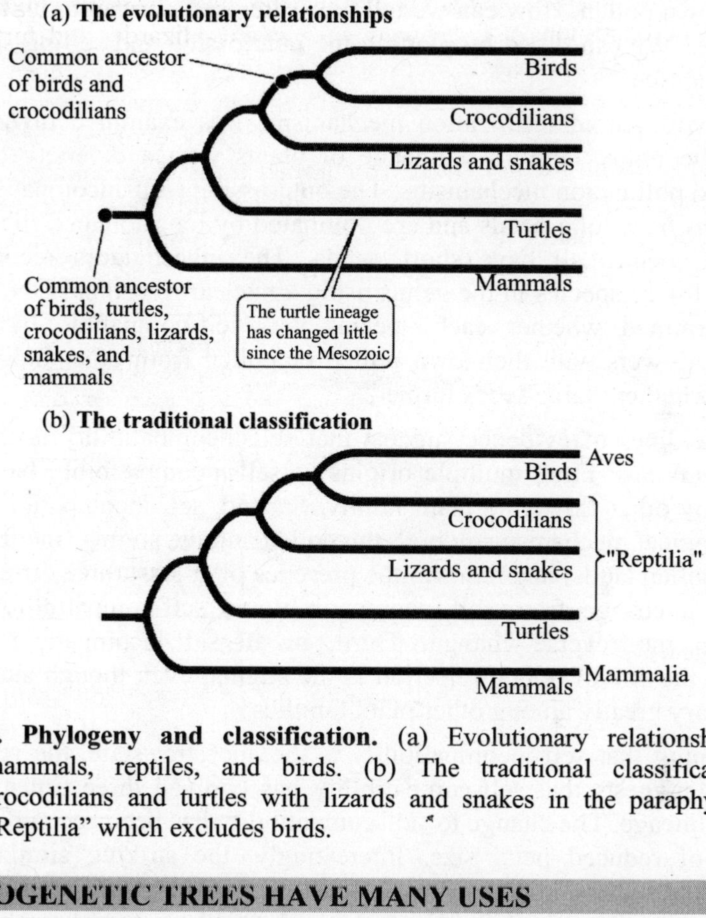

Figure 5.9. Phylogeny and classification. (a) Evolutionary relationships among mammals, reptiles, and birds. (b) The traditional classification unit's crocodilians and turtles with lizards and snakes in the paraphyletic taxon "Reptilia" which excludes birds.

5.6 PHYLOGENETIC TREES HAVE MANY USES

Phylogenies are usually reconstructed as part of an effort to determine evolutionary relationships among organisms. Information about these relationships is useful to scientists investigating a wide variety of biological questions. Many biological statements are really phylogenetic assertions. Any claim of an association between a trait and a group of organisms is actually a statement about when during the history of the lineage the trait first arose and about the maintenance of the trait since its first appearance. For example, the statement that the cytoskeleton is a trait possessed by all eukaryotes is an assertion that the cytoskeleton is an ancestral trait that has been maintained during the subsequent evolution of all surviving eukaryote lineages. In this section, we will illustrate how phylogenetic trees can be used to determine how many times a particular trait may have arisen during evolution, to assess when lineages split, and to explain how evolutionary radiations came about.

5.6.1 How many times has a trait evolved?

Most flowering plants reproduce by mating with another individual—called *outcrossing*—and have mechanisms to prevent self-fertilization. Individuals of some species, however, are *self-compatible*; that is, they can fertilize themselves with their own pollen. How can we tell how often self-compatibility has evolved in a lineage? We can do so by plotting the outcrossing and selfing species on a phylogenetic tree.

The evolution of fertilization mechanisms was examined in *Linanthus* (a genus in the phlox family), a lineage of plants with a diversity of breeding systems and pollination mechanisms. The outcrossing (self-incompatible) species of *Linanthus* have long petals and are pollinated by long-tongued flies. The self-compatible species all have short petals. The investigators reconstructed a phylogeny for 12 species in the genus using a nuclear ribosomal DNA sequence. They determined whether each species was self-compatible by artificially pollinating flowers with their own pollen or pollen from other individuals and observing whether viable seeds formed.

Several lines of evidence suggest that self-incompatibility is the ancestral state in *Linanthus*. First, multiple origins of self-incompatibility have not been found in any other flowering plant family. Second, self-incompatibility depends on physiological mechanisms in both the pollen and the stigma (the female organ on which pollen lands) and requires the presence of at least three different alleles. Therefore, a change from self-incompatibility to self-compatibility would be easier than the reverse change. Third, in all self-incompatible species of *Linanthus*, the site of pollen rejection is the stigma, even though sites of pollen rejection vary greatly among other plant families.

Assuming that self-incompatibility is the ancestral state, the reconstructed phylogeny suggests that self-compatibility has evolved three times within this *Linanthus* lineage. The change to self-compatibility has been accompanied by the evolution of reduced petal size. Interestingly, the striking similarity of the flowers in the self-compatible groups led to their being classified as members of a single species. The phylogenetic analysis shows them to be members of three distinct lineages.

5.6.2 When did lineages split?

How phylogenetic analyses can help us determine when lineages split is illustrated by studies of characiform fishes. The approximately 1,400 species of these freshwater fishes, which are found in both South America (1,200 species) and Africa (200 species), vary greatly in size, shape, and diet. Analyses of mitochondrial DNA and anatomy suggest that the characiform fishes are monophyletic. Analyses using slowly evolving rRNA genes identified three lineages with closely related species on both sides of the Atlantic Ocean. Since all these species live only in fresh water, dispersal across the Atlantic Ocean is unlikely. The genetic differences between the rRNA of the African and the South

American species in all three lineages are great enough to be consistent with a split caused by the separation of Africa from South America, which is believed to have happened about 90 million years ago.

5.7 GENOMES AND THEIR EVOLUTION

An organism's **genome** is the full set of genes that it contains. Most of the genes of eukaryotic organisms are found on chromosomes in the nucleus, but genes are also present in plastids (such as chloroplasts) and mitochondria. In organisms that reproduce sexually, both males and females transmit nuclear genes, but mitochondrial and chloroplast genes usually are transmitted only via the cytoplasm of eggs.

In sexually reproducing organisms, meiosis and fertilization shuffle genes in every generation, and individual gametes transmit partly randomized subsets of parental genes to offspring. For this repeated shuffling to be successful, genes must be able to operate in a wide variety of genetic backgrounds. Closely linked genes, however, are likely to remain together, which is especially important for developmental genes. Many genes that govern developmental processes are closely linked on chromosomes and are inherited together. A gene will not be passed on to successive generations unless the individual with the gene survives and reproduces. Therefore, the capacity to cooperate with different combinations of other genes is likely to increase a gene's probability of transmission. For this reason, it is useful to view the genes of an individual as interacting members of a group, among which there are divisions of labor, but also strong interdependencies. Students of genomic evolution look at the genome of an organism as an integrated whole, asking questions such as,

How do proteins acquire new functions? Why are the genomes of different organisms so variable in size? How has enlargement of genomes been accomplished? Now we will discuss to these questions in the upcoming section of the book, but to provide some needed background, we should keep in mind: molecular evolution which has already been discussed in previous sections of the book in detail.

5.7.1 Proteins Acquire New Functions

Evolution as we know it would not have been possible if proteins were unable to change their functional roles. The earliest forms of life must have had very few genes and proteins. Because much evidence indicates that all living organisms arose from a single ancestral lineage, the many thousands of different functional genes in modern organisms must have arisen from those few ancestral genes. How as this happened?

Proteins may acquire new functions via gene duplication

Gene duplication appears to be the most important process that enables proteins to acquire new functions. When a gene is duplicated, one copy of that gene is potentially freed from having to perform its original function; the copy is redundant if the original gene is still producing the original protein. Therefore, gene duplication may allow the evolution of entirely novel protein functions without impairing cell functioning. Gene duplication may involve part of a gene, a single gene, a portion of a chromosome, an entire chromosome, or the whole genome. Duplication of the entire genome (polyploidy) has been important in speciation, especially in plants. Autopolyploid individuals are usually viable because all of their chromosomes are duplicated. Thus, they avoid imbalances in gene expression. We will discuss gene duplication in greater detail later in this section, but proteins can also acquire new functions in other ways.

Physiological changes may lead to the evolution of new functions for a protein

Lysozyme is an enzyme found in almost all animals. It is produced in the tears, saliva, milk of mammals and in the whites of bird eggs. Lysozyme digests the cell walls of bacteria, rupturing and killing them. As a result, lysozyme plays an important role as a first line of defense against invading bacteria. All animals defend themselves against bacteria by digesting them, which is probably why most animals have lysozyme. Some animals, however, also use lysozyme in the digestion of food.

Among mammals, a mode of digestion called *foregut fermentation* has evolved twice. In animals with this mode of digestion, the foregut—the posterior esophagus and/or the stomach—has been converted into a chamber in which bacteria break down ingested plant matter by fermentation. Foregut fermenters can obtain nutrients from the otherwise indigestible cellulose that makes up a large proportion of the plant body. Foregut fermentation evolved independently in ruminants (a group of hoofed mammals that includes cows) and in certain leaf-eating monkeys, such as langurs. We know that these evolutionary events were independent because both langurs and ruminants have close relatives that are not foregut fermenters. In both foregut-fermenting lineages, the enzyme lysozyme has been modified to play a new, nondefensive role. Lysozyme ruptures some of the bacteria that live in the foregut, releasing nutrients that the mammal absorbs. How many changes in the lysozyme molecule allowed it to function amid the digestive enzymes and acidic conditions of the mammalian foregut? To answer this question, molecular evolutionists compared the amino acid sequences of lysozyme in foregut fermenters and in several of their nonfermenting relatives. They determined which amino acids differed and which were shared among the species (Table 5.2).

Finally, they compared the pattern of these changes with the phylogenetic relationships among the species that had been established based on fossils and current morphology. The most striking finding is that amino acid changes in lysozyme, which occurred in the absence of gene duplication, happened about twice as rapidly in the lineage leading to langurs as in any other primate lineage. This high rate of substitution shows that lysozyme went through a period of rapid change in adapting to the stomachs of langurs. The lysozymes of langurs and cows share five amino acid substitutions, all of which lie on the surface of the lysozyme molecule, well away from the active site. Several of the shared substitutions involve changes from arginine to lysine, which makes the proteins more resistant to attack by the pancreatic enzyme trypsin. By understanding the functional significance of amino acid substitutions, molecular evolutionists can explain the observed changes in amino acid sequences in terms of changes in the functioning of the protein.

Table 5.2. Showing similarity matrix for lysozyme in mammals. Shown the diagonal lines is the amino acid sequence difference between the two species being compared; below the line is the number of sequences uniquely shared by the two species. Asterisks (*) indicate foregut fermenting species.

Similarity Matrix for Lysozyme in Mammals

SPECIES	LANGUR	BABOON	HUMAN	RAT	COW	HORSE
Langur*		14	18	38	32	65
Baboon	0		14	33	39	65
Human	0	1		37	41	64
Rat	0	1	0		55	64
Cow*	5	0	0	0		71
Horse	0	0	0	0	1	

A large body of fossil, morphological, and physiological evidence shows that langurs and cows do not share a recent common ancestor. However, langur and ruminant lysozymes share many amino acid residues that neither animal shares with the lysozymes of their own closer relatives. The lysozymes of these two animals have converged on a similar sequence despite having very different ancestry; in other words, they are homoplasies. The amino acid residues they share give these lysozymes the ability to lyse the bacteria that ferment leaves in the foregut.

An even more remarkable story emerges if we look at lysozyme in the crop of the hoatzin, a leaf-eating South American cuckoo, the only known avian foregut fermenter. Many birds have an enlarged esophageal chamber called a *crop*. Hoatzins have a crop that contains bacteria and acts as a fermenting chamber. Many of the amino acid changes that occurred in the adaptation of hoatzin crop lysozyme are identical to the changes that evolved in ruminants and langurs. Thus, even though the hoatzin and the foregut-fermenting mammals have been evolving independently from one another for more than 100 million years, they have each evolved a similar molecule that enables them to recover nutrients from their fermenting bacteria in a highly acidic environment. The lysozyme story also illustrates why inferring phylogenies from data on single molecules can be very misleading. Identical molecular changes in this protein are not evidence of common descent.

5.7.2 The Evolution of Genome Size

We can now consider what determines the sizes of the genomes of different organisms. As organisms evolved to become more complex, how did the number of functional genes increase so that the organisms could carry out the greater variety of metabolic activities associated with that complexity?

Complex organisms have more DNA than do simpler ones

Genome size varies tremendously among organisms. The first pattern to be detected was that genome sizes are generally correlated with organisms' complexity. The genome of *Mycoplasma genitalium*, the simplest known prokaryote, has only 470 genes. *Rickettsia prowazekii*, the prokaryote that causes typhus, has 634 genes. *Homo sapiens*, on the other hand, has about 21,000 protein coding genes. Figure 5.10 shows the relative sizes of several prokaryotic and eukaryotic genomes. It is not surprising that more complex genetic instructions are needed for building and maintaining a large, complex organism than a small, simple one. What is surprising is that some organisms, such as lungfishes, some salamanders, and lilies, have about 40 times as much DNA as humans do. Clearly, a lungfish or a lily is not 40 times as complex as a human.

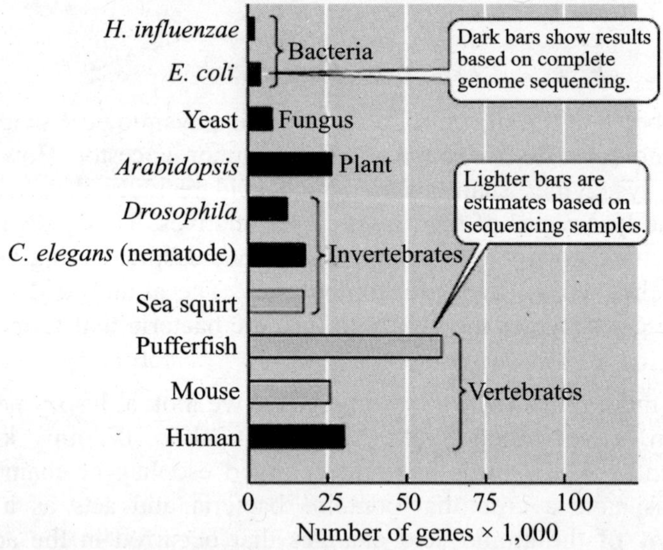

Figure 5.10. Complex organisms have more genes than simpler organisms. Genome sizes have been measured or estimated in a variety of organisms ranging from single-celled prokaryotes to vertebrates.

Why does genome size vary so much? Some of the apparent variation in genome size disappears when we compare the portion of DNA that actually encodes RNAs or proteins. The size of the coding genome of organisms varies in a way that makes sense. Eukaryotes have more coding DNA than prokaryotes; plants have more coding DNA than single-celled organisms; invertebrates with wings, legs, and eyes have more coding DNA than nematodes; and vertebrates have more coding DNA than invertebrates. The organisms with the largest amount of nuclear DNA (some ferns and flowering plants) have 80,000 times as much as the simplest organisms, but no species has more than 20 times as many protein-coding genes as a bacterium. Therefore, most of the variation in genome size lies not in the number of functional genes, but in the amount of noncoding DNA (Figure 5.11).

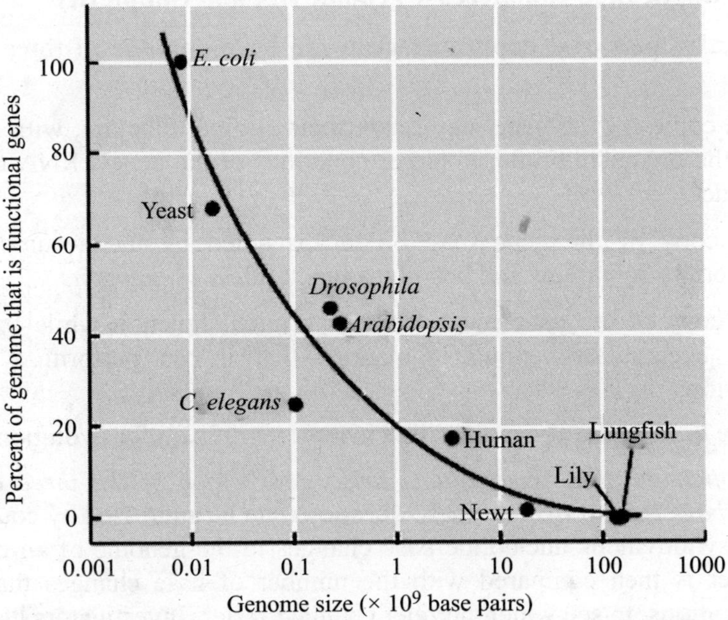

Figure 5.11. A large proportion of DNA is noncoding. Most of the DNA of bacteria and yeasts encodes RNAs or proteins, but most of the DNA of more complex organisms is noncoding. Most noncoding DNA is probably non-functional.

What maintains such large quantities of noncoding DNA in the cells of most organisms? Does this noncoding DNA have a function, or is it "junk"?

Most of this DNA appears to be non-functional. Much of it may consist of pseudogenes that are simply carried in the genome because the cost of doing so is very small. Some of it consists of parasitic transposable elements that spread through populations because they reproduce faster than the host genome. Investigators can use one type of transposable element, retrotransposons, to estimate the rates at which species lose DNA.

Retrotransposons copy themselves with the aid of RNA. The most common type of retrotransposon carries duplicated sequences at each end, called long terminal repeats (LTRs). Occasionally, LTRs join together in the host genome, at which time the DNA between them is excised. When this happens, one of the

LTRs is left behind. The number of such "orphaned" LTRs in a genome is a measure of how many retrotransposons have been lost.

By comparing the number of LTRs in the genomes of Hawaiian crickets of the genus *Laupala* and those of fruitflies (*Drosophila*), investigators found that *Laupala* loses DNA more than 40 times more slowly than *Drosophila*. As a result, the genome of *Laupala* is 11 times larger than that of *Drosophila*. Why species differ so greatly in the rate at which they lose DNA is not understood.

5.7.3 Gene duplication can increase genome size and complexity

The identical copies of a duplicated gene can have any one of three different fates:

- Both copies of the gene may retain their original function, with the result that the organism produces larger quantities of the gene's RNA or protein product.
- One copy of the gene may be incapacitated by the accumulation of deleterious mutations and become a functionless *pseudogene*.
- One copy of the gene may retain its original function while the second copy accumulates enough mutations that it can perform a different function.

It is the third of the above fates that is most significant for evolution.

How often do gene duplications arise, and which of the three outcomes described above is most likely? These questions can be addressed by counting the number of synonymous nucleotide base changes in the genome of an organism. This number is then compared with the number of base changes that caused protein alterations, to see which number changed faster. Investigators have found that rates of gene duplication are fast enough for a yeast or *Drosophila* population to acquire several hundred duplicate genes over the course of a million years. They also found that most of the duplicated genes in these organisms are very young. Extra genes typically are lost from a genome within 10 million years (which is rapid on an evolutionary time scale).

Although extra genes usually disappear rapidly, some duplication events lead to the evolution of genes with new functions. Several successive rounds of duplication and mutation may result in a **gene family**, a group of homologous genes with related functions, often arrayed in tandem along a chromosome. An example of this process is provided by the globin gene family.

The globins were among the first proteins to be sequenced and compared. Comparisons of their amino acid sequences strongly suggest that the different globins arose via gene duplications. How long the globins have been evolving separately can also be inferred by comparing their amino acid sequences. The greater the number of amino acid differences between two globins, the further back in time was their most recent common ancestor.

Hemoglobin, a tetramer consisting of two α-globin chains and two β-globin chains, carries oxygen in blood. Myoglobin, a monomer, is the primary oxygen storage protein in muscle. Myoglobin's affinity for O_2 is much higher than that of hemoglobin. In contrast, hemoglobin evolved to be more diversified in its role. Hemoglobin binds O_2 in the lungs or gills, where the O_2 concentration is relatively high, transports it to deep body tissues, where the O_2 concentration is low, and releases it in those areas. With its more complex tetrameric structure, hemoglobin is able to carry four molecules of O_2, as well as hydrogen ions and carbon dioxide, in the blood.

To estimate the time of the globin gene duplication that gave rise to the α- and β-globin gene clusters, we can create a **gene tree** based on the estimated number of base substitutions necessary to account for the observed amino acid differences between the globins. Based on this gene tree, and assuming that the rate of amino acid substitution has been relatively constant since then—about 100 substitutions per 500 million years—the two globin gene clusters are estimated to have split about 450 million years ago (Figure 5.12).

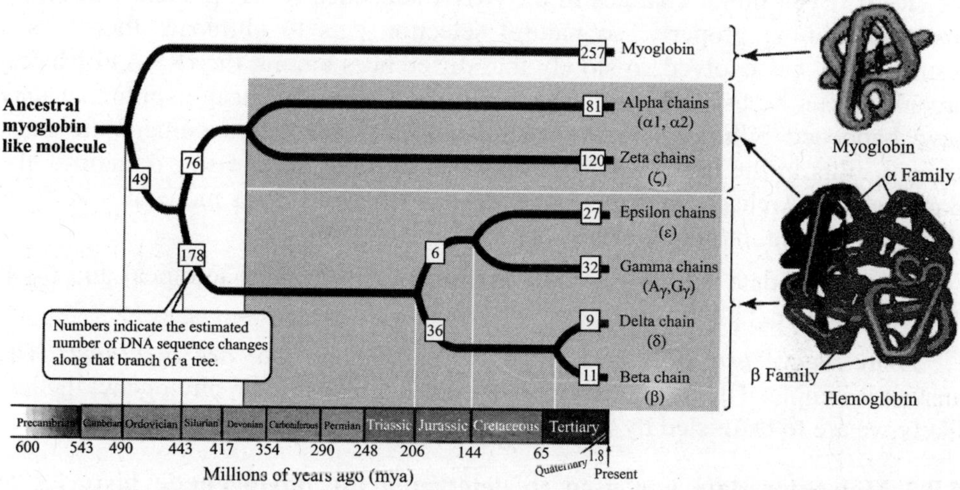

Figure 5.12. A globin gene family tree. This gene tree suggests that the α-globin and β-globin gene clusters diverged about 450 mya, at about the time of the origin of the vertebrates.

5.8 THE USES OF MOLECULAR GENOMIC INFORMATION

Information about the genomes of organisms enables biologists to investigate a variety of biological problems in new and powerful ways. Information about the rates at which molecules have evolved is used in reconstructing phylogenetic trees.

Molecular genomic information can also be used to determine the phylogenies of genes, which often differ from those of organisms because genes can be transferred from one evolutionary lineage to another, a phenomenon known as *lateral gene transfer*.

Recent evidence suggests that such lateral transfers may have occurred repeatedly during the evolution of plants. And, as we discussed at the beginning of this chapter, genomic information has powerful medical applications.

5.8.1 Molecular information is used to reconstruct phylogenies

By comparing the structures of molecules from different species, we both gain insights into how those molecules function and acquire a tool for inferring phylogenies. Molecules that have evolved slowly can be used to reconstruct relationships among organisms that diverged long ago. Molecules that have evolved rapidly are useful for studying organisms that share more recent common ancestors.

To investigate the evolutionary relationships of all existing organisms, we choose molecules that all organisms possess, such as ribosomal RNA. Equally important to our choice are strong functional constraints, such as those that exist for rRNA. Even minor changes in the rRNA sequence would prevent ribosomes from functioning properly, so natural selection acts to eliminate them. As a result, rRNA has evolved so slowly that differences among the rRNAs of living organisms can be used to estimate the relative timing of lineage splits that may have happened billions of years ago. No fossils exist to document the most ancient splits in the lineages of life on Earth into the three major domains—the Bacteria, the Archaea, and the Eukarya—so we depend on molecules such as rRNA for insight into these events.

Molecular data are also used in combination with morphological and fossil data to reconstruct phylogenies.

Why do we use molecules when morphological data are available? The answer is simple: The more characters we use to reconstruct a phylogeny, the less likely we are to be misled by evolutionary reversals or convergent evolution.

5.8.2 Molecular data are used to determine the phylogenetic histories of genes

A gene tree shows the evolutionary relationships of members of a gene family, just as a phylogenetic tree depicts the evolutionary relationships of members of a lineage. All of the genes of a particular family have similar sequences because they have a common ancestry.

Genes found in different organisms that arose from a single gene in their common ancestor are called **orthologs**.

Genes that are related through gene duplication events in a single lineage are called **paralogs**.

5.8.3 Molecular information provides new ways to combat diseases

The effort to determine the genomes of *Anopheles* and *Plasmodium* has already had medical benefits. Transgenic mosquitoes have been engineered to express an anti-*Plasmodium* molecule that makes them inefficient vectors of malaria in the laboratory. This advance would not have been possible without knowledge of the parasite's genome. In addition to genetic information, success in controlling malaria will require detailed knowledge of the behavior, ecology, and evolution of the many species of mosquitoes that transmit malaria in different parts of the world.

Information provided by the genomic sequence of *Treponema pallidum*, the bacterium that causes syphilis, is being used to help develop a vaccine against this sexually transmitted disease. Syphilis may have been introduced into Europe from the Americas by members of Columbus's expedition, and the disease spread across the world during the sixteenth-century age of exploration. Effective therapies have been available since the discovery of penicillin in the mid-twentieth century, but syphilis remains a serious global health problem.

Treponema pallidum has been a difficult organism to study because it cannot be grown outside a mammalian host. The genome sequence of *T. pallidum* may provide the information necessary for the development of a culture medium. In addition, the sequence revealed that *T. pallidum* contains a family of paralogous genes (*tprA–L*) that encode proteins of the bacterial outer membrane. The identification of this gene family is guiding the search for targets for a vaccine against the disease.

Molecular data cannot solve all disease problems

Molecular data are powerful tools in our struggle with diseases, but the AIDS epidemic shows that they cannot solve all medical problems. The agent that causes AIDS is a retrovirus (HIV) that uses RNA as its genetic material. The central core of the virus contains two identical copies of the RNA genome as well as three enzymes needed to carry out the viral life cycle: reverse transcriptase, integrase, and a protease.

The highly active antiretroviral therapy (HAART) most commonly used against AIDS today employs a combination of protease inhibitors and reverse transcriptase inhibitors, each of which blocks a different stage in the viral life cycle. Unfortunately, resistant strains of HIV develop in the blood of most patients that receive HAART. HIV's mutation rate is very high, and since it has no repair enzymes, a wrong base is inserted at about one out of every 8,000 nucleotide positions. The result is that, on average, a new mutant is generated every time HIV replicates its genome—in effect, no two viruses are identical!

Although biomedical scientists are now armed with detailed knowledge of the molecules that govern the different stages in the viral life cycle, HIV's extremely high mutation rate has prevented them from coming up with drugs to which the virus cannot rapidly adapt. In the absence of such drugs, reducing both the incidence and severity of HIV will require changes in human sexual behavior.

The development of molecular methods, the sequencing of the genomes of an increasing number of species, and powerful computers have ushered in a new era in the scientific study of the diversity of life. Scientific understanding of the evolutionary patterns of life on Earth and how the agents of evolution have governed these patterns is advancing more rapidly than at any prior time during the study of evolution. The range of data used by systematists is likely to continue to increase because, using modern chemical, biochemical, and microscopic methods, we will be able to measure more traits of organisms than we could previously.

By combining molecular data with information from the fossil record (which is also increasing rapidly), biologists are developing a comprehensive picture of the evolution of life on earth.

■■■■

Chapter 6

VERTEBRATE EVOLUTION AND DIVERSITY

6.1 Invertebrate Chordates and the Origin of Vertebrates

6.1.1 Four anatomical features characterize phylum Chordata

6.1.2 Invertebrate chordates provides clues to the origin of vertebrates

6.2 Introduction to the Vertebrates

6.2.1 Characteristics of Subphylum Vertebrates

6.2.2 Overview of vertebrate diversity

6.3 Superclass Agnatha: Jawless Vertebrates

6.3.1 Lampreys and hagfish are the only extant agnathans

6.4 Superclass Gnathostomata I: The Fishes

6.4.1 Evolution of vertebrate jaws from skeleton support of the pharyngeal slits

6.4.2 Class Chondrichthyes: A cartilaginous endoskeleton

6.4.3 Class Osteichthyes: A bony endoskeleton

6.5 Superclass Gnathostomata II: The Tetrapods

6.5.1 Amphibians are the oldest class of tetrapods

6.5.2 Evolution of the amniotic egg expanded the success of vertebrates on land.

6.5.3 Reptilian evolution

6.5.4 Evolution of birds

6.5.5 Mammalian evolution

6.6 Primates and the Evolution of *Homo sapiens*

6.6.1 Primate evolution provides a context for understanding human origins

6.6.2 Humanity: a very young twig on the vertebrate tree

6.6.3 Four main classes of change occurred during hominin evolution

6.6.4 Occurrence of hominids to be classified as genus *Homo*

6.6.5 The emergence of anatomically modern humans: *Homo sapiens*

6.7 Cultural Evolution

6.1 INVERTEBRATE CHORDATES AND THE ORIGIN OF VERTEBRATES

Based on certain similarities in early embryonic development, chordates are grouped as deuterostomes along with the echinoderms. The vertebrates make up one subphylum within the phylum Chordata. Chordates also include two subphyla of invertebrates, the urochoradates and the cephalochordates.

6.1.1 Four anatomical features characterize phylum Chordata

Although chordates vary widely in appearance, they are distinguished as a phylum by the presence of four anatomical structures that appear at some point during the animals life time, often only during embryonic development. These four chordate characteristics are: a notochord; a dorsal, hollow nerve cord; pharyngeal slits; and a muscular, postnatal tail (Figure 6.1).

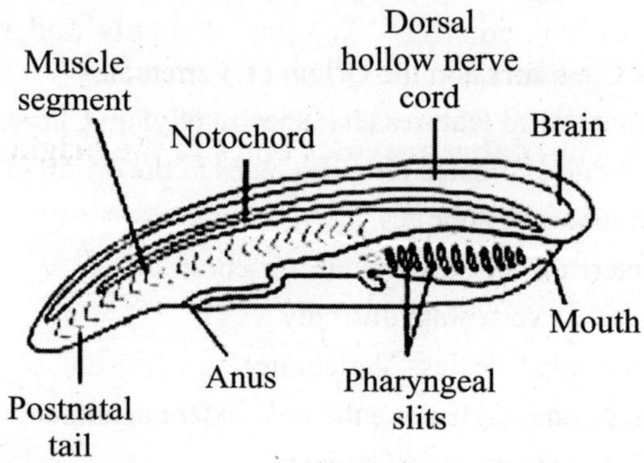

Figure 6.1. Chordate Characteristics

1. Notochord

Chordates are named after a skeletal structure, the notochord, present in all chordate embryos. The notochord is a longitudinal, flexible rod located between the digestive tube and the nerve cord. Composed of large, fluid filled cells encased in fairly stiff, fibrous tissue, it provides skeleton support throughout most of the length of the animal. The notochord persists in the adults of some invertebrate chordates and primitive vertebrates. However, in most vertebrates a more complex, jointed skeleton develops, and the adult retains only remnants of embryonic notochord – as gelatanious material of the discs between vertebrae of humans, for example.

2. Dorsal, Hollow Nerve Cord

The nerve cord of a chordate embryo develops from a plate of ectoderm that rolls into a tube located dorsal to the notochord. The result is a dorsal, hollow nerve cord unique to chordates. Other animal phyla have solid nerve cords, usually ventrally located. The nerve cord of a chordate embryo develops into the central nervous system: the brain and the spinal cord.

3. Pharyngeal Slits

The digestive tube of chordates extends from the mouth to the anus. The region just posterior to the mouth is the pharynx, which opens to the outside of the animal through several pairs of slits. These pharyngeal slits allow water that enters the mouth to exit without continuing through the entire digestive tract. The pharyngeal gill slits function as suspension feeding devices in many invertebrate chordates. The slit and structures supporting them have become modified for gas exchange (in aquatic vertebrates), jaw support, hearing, and other functions during vertebrate evolution.

4. Muscular, Postnatal Tail

Most chordates have a tail extending posterior to the anus. By contrast, most nonchordates have a digestive tract that extends nearly the whole length of the body. The chordate tail contains skeleton elements and muscles and provides much of the propulsive force in many aquatic species

6.1.2 Invertebrate chordates provides clues to the origin of vertebrates

Members of the two subphyla of invertebrate chordates, Urochordata and Cephalochordata, illustrate the chordate body plan in its most "stripped-down" versions without the additional features that evolved in vertebrates. The study of urochordates and cephalochordates also provides clue to the origin of vertebrates.

Subphylum Urochordata

Urochordates are commonly called tunicates. Most tunicates are sessile marine animals that adhere to rocks, docks, and boats (Figure 6.2). Others are planktonic. Some species are colonial. Seawater enters the animal through an incurrent siphon, passes through the pharyngeal slits into a chamber called the atrium, and exist through an excurrent siphon, or atriopore (Figure 6.3). The food filtered from this water current by a mucus net is passed by cilia into the intestine. The anus empties into the excurrent siphon. The entire animal is cloaked in a tunic made of cellulose-like carbohydrate. Because they shoot a jet of water through the excurrent siphon when molested, tunicates are also called as sea squirts.

Figure 6.2. An adult sessile tunicate

The adult tunicate scarcely resembles a chordate. It displays no trace of a notochord, nor is there a nerve cord or tail. Only the pharyngeal slits suggest a link to other chordates. But all four chordate trademarks are manifest in the larval form of some groups of tunicates (Figure 6.4). The larva swims until it attaches by its head to a surface and undergoes metamorphosis, during which most of its chordate characteristics disappear. In at least some of the tunicates, if a single gene *Manx* is turned off during the development, a normally tailed tunicate larva will be tailless. Expression of *Manx* is also essential for notochord and dorsal nerve cord development. This supports strongly the hypothesis that a relatively small numbers of genes regulating development may have been associated with the evolution of some basic features of an animal body plan – in this case, the suite of characteristics that define all chordates.

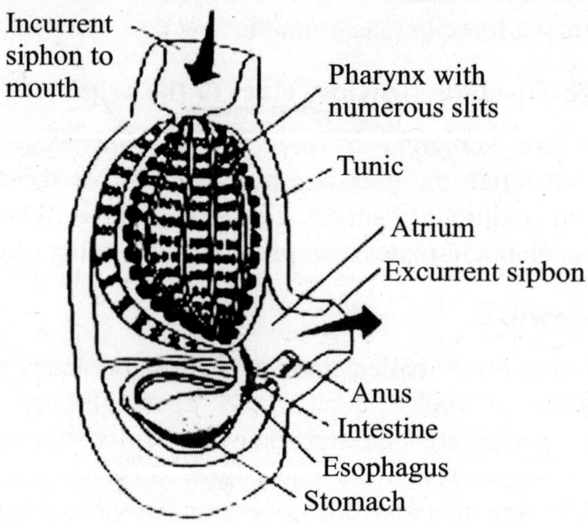

Figure 6.3. The adult tunicate morphology showing various body parts but other chordate characteristics are not evident.

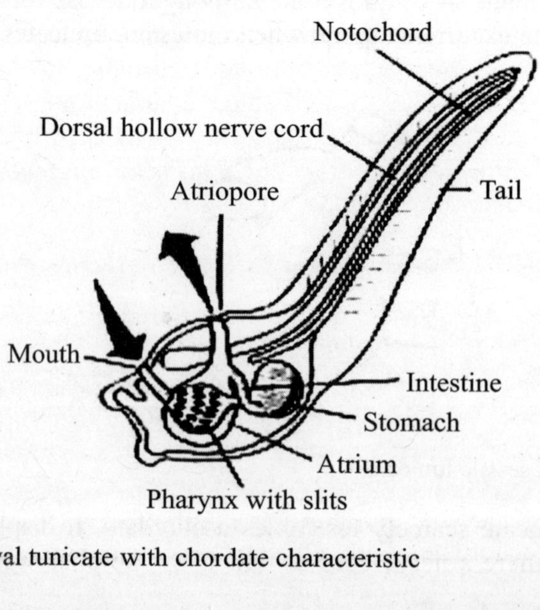

Figure 6.4. The larval tunicate with chordate characteristic

Subphylum Cephalochordata

They are known as lancelets because of their bladelike shape, cephalochordates closely resemble the idealized chordate in Figure 6.1. The notochord; dorsal, hollow nerve cord; numerous gill slits; and postnatal tail all persist into the adult stage (Figure 6.5). Small marine animals only a few centimeters long, lancelets wriggle backward into the sand, leaving only their anterior end exposed. A mucus net secreted across the pharyngeal slits traps tiny food particles from seawater drawn into the mouth by ciliary pumping. The water exits through the slits, and the trapped food passes down the digestive tube.

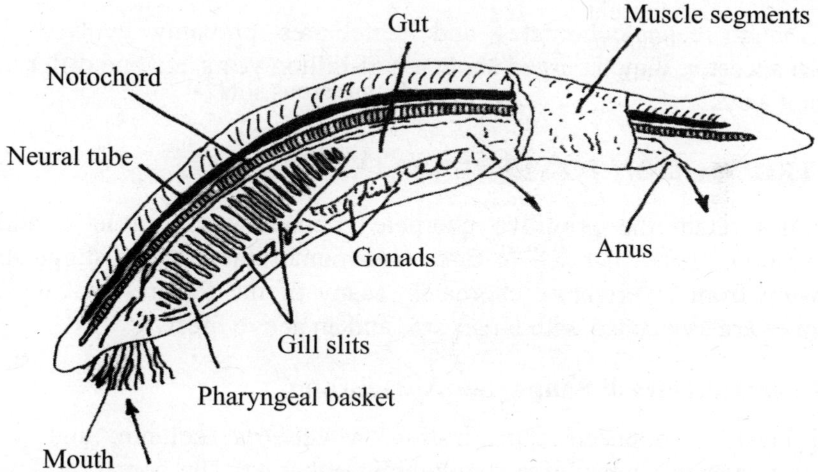

Figure 6.5. Cephlochordata: The adult Branchiostoma displaying all four chordate characteristics.

A lancelet frequently leaves its burrow and swims to a new location. Though feeble swimmers, these invertebrate chordates display, in simple form, the swimming mechanism of fishes. Coordinated contraction of muscles serially arranged along the sides of notochord flexes the notochord, producing side-to-side undulations that thrust the body forward. This serial musculature is evidence of the lancelets segmentation.

Relationship between Invertebrate chordate and Vertebrates

Paleontologists have found fossilized invertebrates resembling cephalochordates in the Burgess Shale of British Columbia, Canada. Those fossils are about 545 million years old, about 50 million years older than the oldest known vertebrates. The record of the rock is too incomplete for us to retrace the origin of the earliest vertebrates from invertebrate ancestors, but we can propose reasonable hypotheses about this evolution based on comparative anatomy and embryology. Many biologists think that the ancestor of vertebrates was a suspension-feeder similar to a cephalochordate, with all four basic chordate features. Recent research by molecular systematists supports the hypothesis that cephalochordates are the vertebrate's closest relatives.

Both cephalochordates and vertebrates may have evolved from a common sessile ancestor by paedogenesis, the precious development of sexual maturity in a larva. Cephalochordate appears more akin to an adult urochordate. Changes in gene controlling development can alter the timing of developmental events, such as the maturation of gonads. Perhaps such a change occurred in the ancestor of cephalochordates and vertebrates, causing the gonads to mature in swimming larvae before the onset of metamorphosis to the sessile adult form. If reproducing larvae were very successful, natural selection may have reinforced paedogenesis and eliminated metamorphosis.

Although cephalochordates and vertebrates probably evolved from a common ancestor, they diverged about a half-billion years ago and differ in many important ways.

6.2 INTRODUCTION TO VERTEBRATES

Vertebrates retain the primitive chordate characteristics but have additional specialization, shared derived features (synapomorphies) that distinguished the subphylum from invertebrate chordates. Many of the features that distinguish vertebrates are associated with larger size and an active lifestyle.

6.2.1 Characteristics of Subphylum Vertebrates

Neural crest, pronounced cephalization, a vertebral column, and a closed circulatory system characterize subphylum vertebrata. The dorsal, hollow nerve cord found in all chordata develops when the edges of an ectodermal plate on the surface of the embryo roll to form the neural tube. In vertebrates, a group of embryonic cells called the neural crests forms near the dorsal margins of the closing neural tube (Figure 6.6). A distinctive feature of vertebrates, the neural crest, contributes to the formation of certain skeletal elements, such as some of the bone and cartilage of the cranium, and many other structures that distinguish vertebrates from other chordates. The vertebrate cranium and brain (which is the enlarged anterior end of the dorsal hollow nerve cord), along with the eyes, ears, and nose, are evidence of an important evolutionary feature of vertebrates – a high degree of cephalization, the concentration of sensory and neural equipment in the head.

The cranium and vertebral column, surrounding and protecting the nerve cord, are parts of the vertebrate axial skeleton, the main support structure for the axis, or central trunk, of the body. The axial skeleton helps make large body size and strong, fast movement possible. The axial skeleton of most vertebrates also includes ribs, which anchor muscles and protect internal organs. Vertebrates also have an appendicular skeleton, supporting their two pairs of appendages (fins, legs, or arms).

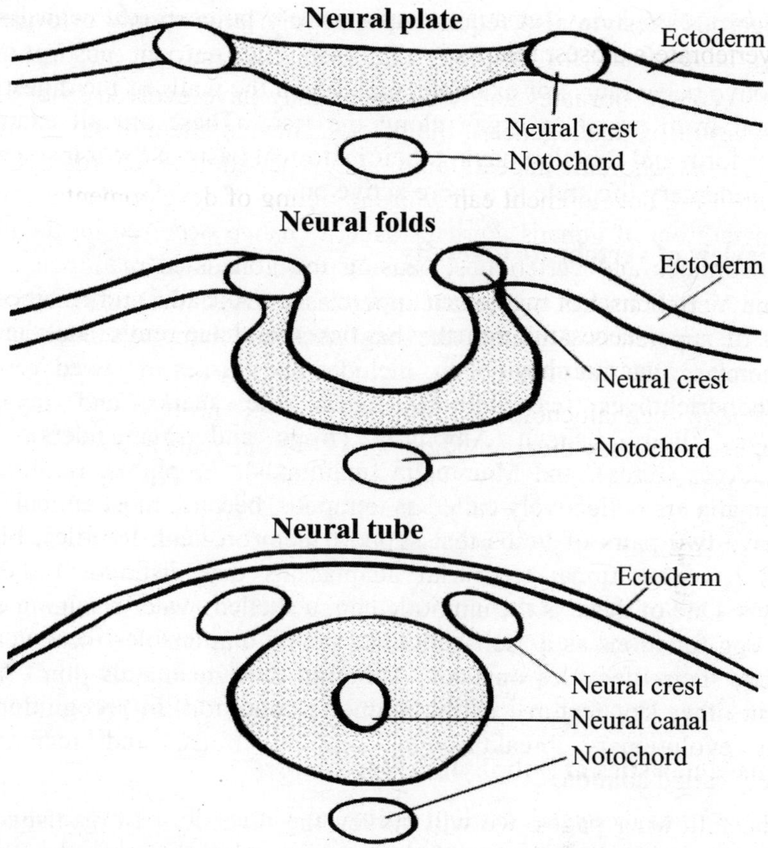

Figure 6.6. The neural crest. The neural crest consists of bilateral bands of cells near the margins of the embryonic folds that meet to form the dorsal, hollow nerve cord (neural tube).

The vertebrate endoskeleton is made of bone or more flexible cartilage, or most commonly some combination of these two materials. Although the skeleton consists mostly of a nonliving matrix, living cells within the skeleton secrete and maintain the materials of the matrix. The endoskeleton of a vertebrate can grow with the animal, unlike the nonliving exoskeleton of arthropods, which must be periodically shed and rebuilt as the animal grows.

When vertebrates move in pursuit of food or when escaping predators, they regenerate their ATP supply mainly by cellular respiration, which consume oxygen. Adaptation of the vertebrates respiratory and circulatory system supports busy mitochondria in muscle cells and other active tissues. Vertebrates have a closed circulatory system, with ventral, chambered heart that pumps blood though arteries to microscopic vessels called capillaries that branch throughout every tissue in body. The blood is oxygenated as it passes through capillaries in gills or lungs.

A vigorous lifestyle also requires a relatively large supply of organic fuel. Vertebrate adaptations for feeding, digestion and nutrient absorption helps support active behaviour. For example, muscles in the walls of the digestive tract propel food from organ to organ along the tract. These are all examples of vertebrates form and function having their historical basis in the transition from a relatively sedentary life style to a more active one.

6.2.2 Overview of vertebrate diversity

Subphylum Vertebrata has two extant superclasses: Agnatha and Gnathostomata. Members of superclass Agnatha, the hagfishes and lampreys, lack jaws. The other superclass, the Gnathostomata, includes six classes of jawed vertebrates: Class Chondrichthyes (cartilaginous fishes, the sharks and rays; Class Osteichthyes (bony fishes); Amphibia (frogs and salamanders), Reptilia (reptiles), Aves (birds), and Mammalia (mammals). Amphibia, Reptilia, Aves, and Mammalia are collectively called as tetrapods, because most animals in these classes have two pairs of limbs that support them on land. Reptiles, birds, and mammals have additional terrestrial adaptations that distinguish them from amphibians. One of these is the amniotic egg, a shelled, water-retaining egg. The amniotic egg functions as a "self-contained pond" that enables these vertebrates to complete their lifecycles on land. Although most mammals don't lay eggs, they retain other key features of the amniotic condition. In recognition of this important evolutionary breakthrough, reptiles, birds, and mammals are collectively called amniotes.

In the following pages, we will survey the diversity of organisms in each vertebrate class, highlight some of the major evolutionary trends among the vertebrates, and examine some of the taxonomic debates about the vertebrate classes.

6.3 SUPERCLASS AGNATHA: JAWLESS VERTEBRATES

Vertebrates probably originated in the late Precambrian or early Cambrian era. The oldest vertebrate fossils are jawless creatures, members of superclass Agnatha (without jaws). Traces of these early vertebrates are found in Cambrian strata, but most date back to the Ordovician and Silurian periods, about 400 to 500 million years ago. Superclass Agnatha includes extinct fishlike animals called *Ostracoderms* (shell skinned) that were encased in an armor of bony plates. These and other early agnathans were generally small, less than 50 cm in length. Most lacked paired fins and apparently were bottom-dwellers that wiggled along streambeds or the seafloor, but there were also some more active species with paired fins. Their mouths were circular or slitlike openings that lacked jaws. Most agnathans were probably mud-suckers or suspension-feeders that took in sediments or suspended organic debris through their mouths and then passed it through the gill slits, where food was trapped. Thus, the pharyngeal apparatus retained the primitive feeding function, although gills in agnathans were probably also the major sites of gas exchange. The ostracoderms and most other agnathans groups declined and finally disappeared during the Devonian period.

6.3.1 Lampreys and hagfish are the only extant agnathans

About 60 species of jawless vertebrates are alive today in Class Myxini (hagfish) and Class Cephalaspidomorphi (lampreys) (Figure 6.7). In common with many extinct agnathans, lampreys and hagfishes lack paired appendages. In contrast to many ostracoderms, however, they have no external armor. The eel-shaped sea lamprey feeds by clamping its round mouth onto the flank of a live fish, using a rasping tongue to penetrate the skin of its prey, and ingesting the prey's blood. Sea lampreys live as larvae for years in fresh water streams and then migrate to the sea or lakes as they mature into adults. The larva is a suspension-feeder that resembles the lancelets (cephalochordates). Some species of lampreys feeds only as larvae. Following several years in streams, they attain sexual maturity, reproduce, and die within a few days.

Hag fishes superficially resemble lampreys, but they are mainly scavengers rather than blood-suckers or suspension-feeders, and their mouth parts are not adapted for rasping. Some species feed on sick or dead fish, whereas other hagfishes eat marine worms. Hagfishes lack a larval stage and live in salt water.

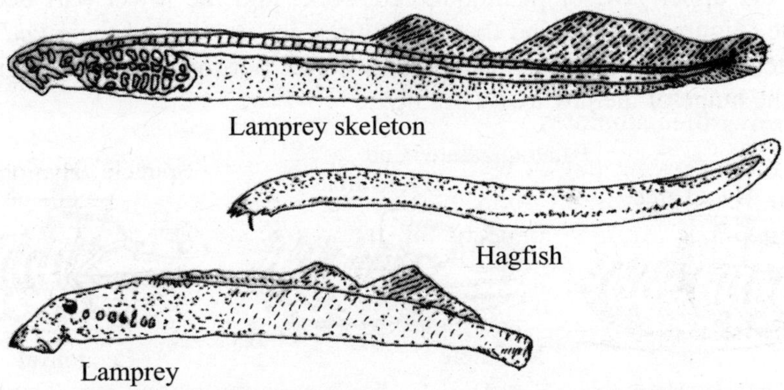

Figure 6.7. A Sea lamprey and Hagfish, jawless vertebrates (Super class Agnatha)

6.4 SUPERCLASS GNATHOSTOMATA I: THE FISHES

During the late Silurian and early Devonian periods, vertebrates with jaws, members of Superclass Gnathostomata (jaw mouth) largely replaced the agnathans. The extant classes of fishes (Chondrichthyes and Osteichthyes) first appeared at this time, along with a group called the placoderms (plate skinned) that has no living descendants. Jawed vertebrates also have two pairs of paired appendages, whereas agnathans either lacked paired appendages or had only a single pair. Judging from studies on extant vertebrates, differential expression of some of the *Hox* genes may have determined whether one or two pairs of paired appendages developed.

The largest placoderms were more than 10 m long, but most were less than 1m long. With two pairs of paired fins and hinged jaws, many species were active predators, capable of chasing prey and biting off chunks of flesh. Thus, these two modifications of the early vertebrate body plan allowed the diversification of both lifestyles and nutrient sources.

6.4.1 Evolution of vertebrate jaws from skeleton support of the pharyngeal slits

The origin of jaws was a major adaptive event in early vertebrate phylogeny. Vertebrate jaws are hinged and work up and down (dorsoventrally). Hinged jaws also evolved in arthropods, but these had a different origin than vertebrate jaws. Arthropod jaws are modified appendages that work from side to side.

Vertebrate jaws evolved from the transformation of pharyngeal gill arches previously used in filter feeding and perhaps respiration as well. These arches are paired on each side and supported by V-shaped hinged structures whose apices point backward. However the fate of anterior pairs of gill arches remains unclear; the arches may have disappeared, incorporated into the base of the cranium, or formed one or more of the mouth structures. Whatever happened, the first gill arch posterior to these anterior pairs changed so that the upper part of the hinge became the upper jaw, or palatoquadrate bone, and the lower part became the mandible (Figure 6.8). Behind these structures, an arch called the *hyoid* incorporated into the complex by contributing its dorsal portion, the hyomandibular, to anchor the hinge of the jaw to the braincase.

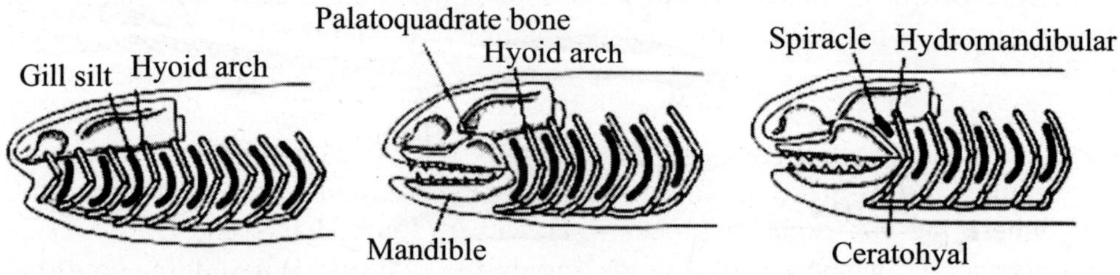

Figure 6.8. Stages of evolution of vertebrate jaws

The origin of the vertebrate jaws from these skeleton parts illustrates a general feature of evolutionary change. New adaptations usually evolve by the modification of existing structures. As a mechanism of adaptation, evolution is limited by the raw material with which it must work; evolution is generally more of a remodeling process than a creative one.

The Devonian period (about 360 to 400 million years ago) has been the age of fishes, when the placoderms and another group of jawed fishes called acanthodians (placed in separate classes) radiated and many new forms evolved in both fresh and salt water. Placoderms and acanthodians dwindled and disappeared almost completely by the beginning of the carboniferous period, about 360 million years ago. Ancestors of the placoderms and acanthodians may also have given rise to sharks (class Chondrichthyes). Another group of large, jawed predators – the bony fishes (class Osteichthyes) diverged from the common ancestors some 425 to 450 million years ago. Sharks and bony fishes have teeth in their jaws, and the teeth are replaced regularly. In contrast, the placoderms and acanthodians had plates of external armor forming the edges of the jaws. Sharks and bony fishes are the reigning vertebrates over the watery two-thirds of the surface of earth.

6.4.2 Class Chondrichthyes: A cartilaginous endoskeleton

The vertebrates of class Chondrichthyes, sharks and their relatives, are called cartilaginous fishes because they have relatively flexible endoskeletons made of cartilage rather than bone. However, in most species, parts of the skeleton are strengthened by calcified granules. There are about 750 extant species in this class. Jaws and paired fins are well developed in the cartilaginous fishes. The largest and most diverse subclass is composed of a few dozen species of unusual fishes called chimaeras, or ratfishes. The cartilaginous skeleton of these fishes is a derived characteristic, not a primitive one; that is, the ancestors of Chondrichthyes had bony skeletons, and the cartilaginous skeleton characteristic of the class evolved secondarily. During the development of most vertebrates, the skeleton is first cartilaginous and then becomes bony (ossified) as hard calcium phosphate matrix replaces the leathery matrix of cartilage. Apparently, some modification in the developmental process of cartilaginous fishes prevents the replacement process.

Most sharks have streamlined bodies (Figure 6.9a). They are swift swimmers, powerful swimming muscles, especially in the caudal (tail) fin, propel them forward. The dorsal fins function mainly as stabilizers, and the paired pectoral (fore) and pelvic (hind) fins provide lift in the water. Although a shark gains additional buoyancy by storing a large amount of oil in its huge liver, the animal is still denser than water, and it sinks if it stops swimming. Continual swimming also ensures that water will flow into the mouth and out through the gills, where the gas exchanges occurs. However, some sharks and many skates and rays spend a good deal of time resting on the seafloor. When doing so, these fishes use muscles of the jaws and pharynx to pump water over the gills.

The largest sharks and rays are suspension-feeders that feed on plankton. Most sharks, however, are carnivores that swallow their prey whole or use their powerful jaws and sharp teeth to tear flesh from animals too large to swallow in one piece. Shark teeth probably evolved from the jagged scales that cover the abrasive skin. The digestive tract of many sharks is proportionately shorter than the digestive tube of many other vertebrates. Within the shark intestine is a spiral valve, a corkscrew-shaped ridge that increases surface area and prolongs the passage of food along the short digestive tract.

Acute senses are adaptations that go along with the active, carnivorous lifestyle of sharks. Sharks have sharp vision but cannot distinguish colours. The nostrils of sharks, like those of most fishes, open into dead-end cups. They function only for olfaction (smelling), not for breathing. Along with eyes and nostrils, the shark head also has a pair of regions in the skin that can detect electrical fields generated by the muscle contractions of nearby animals. Running the length of each flank of the shark is the lateral line system, a row of microscopic organs sensitive to changes in the surrounding water pressure. A primitive feature and present in most species of aquatic vertebrates, the lateral line system enables the animal to detect minor vibrations. Sharks and other fishes have no eardrums, structures in terrestrial vertebrates that convey sound waves traveling through air into the ear toward the auditory organs. Sound reaches the shark through water, and the animal's entire body conveys the sound to the hearing organs of the inner ear.

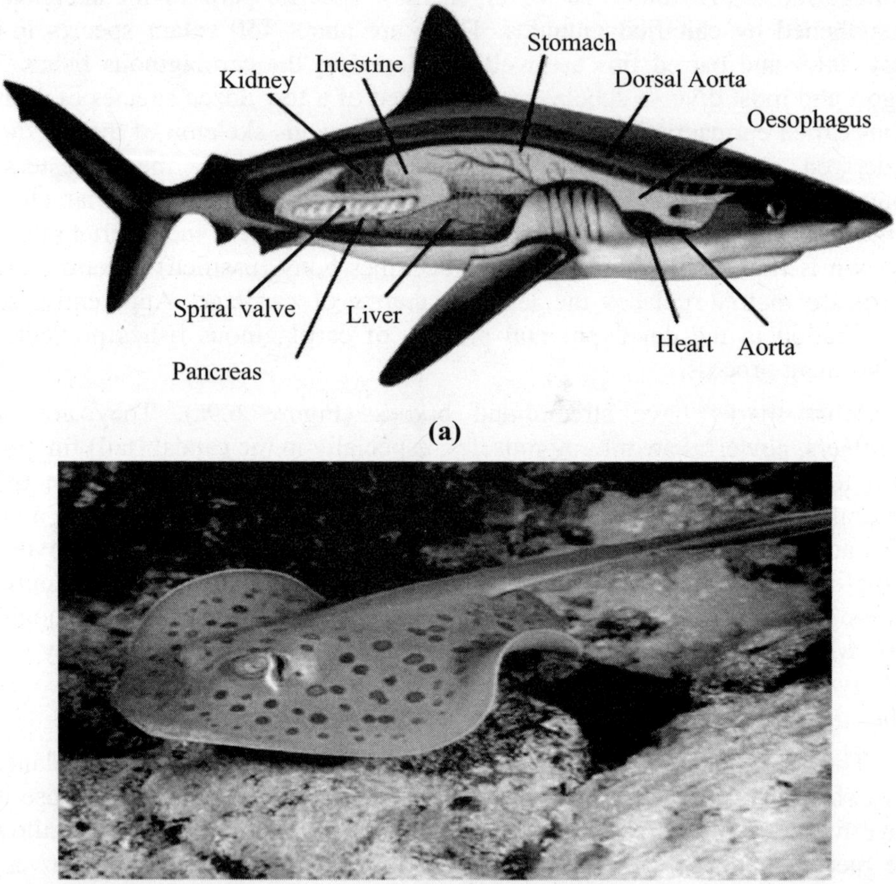

Figure 6.9. Cartilaginous fishes. (a) Morphology of a shark, (b) A blue-spotted stingray fish

Sharks eggs are fertilized internally. The male has a pair of claspers on its pelvic fins that transfer sperm into the reproductive tract of the female. Some species of sharks are oviparous; they lay eggs that hatch outside the mother's body. These sharks release their eggs after encasing them in protective coats.

Other species are ovoviviparous; they retain the fertilized eggs in the oviduct. Nourished by the egg yolk, the embryos develop into young that are born after hatching within the uterus. A few species are viviparous; the young develop within the uterus, nourished prior to birth by nutrients received from the mother's blood through the placenta. The reproductive tract of the shark empties along with the excretory system and digestive tract into the cloaca, a common chamber that expel through a single vent.

Although rays are closely related to sharks, they have adopted a very different lifestyle. Most rays are flattened bottom-dwellers that feed by using their jaws to crush mollusks and crustaceans (Figure 6.9 b). A rays pectoral fins are greatly enlarged and are used to propel the animal through the water. The tail of many rays is whiplike and, in some species, bears venomous barbs that function in defense.

6.4.3 Class Osteichthyes: A bony endoskeleton

Of all vertebrate class, bony fishes (class Osteichthyes) are the most numerous, both in individuals and in species (about 30,000). Ranging in size from about 1 cm to more than 6 m long, bony fishes are abundant in the seas and in nearly every freshwater habitat. Nearly all bony fishes have an endoskeleton with a hard matrix of calcium phosphate. The skin is often covered by flattened bony scales that differ in structure from the toothlike scales of sharks. Glands in the skin of a bony fish secrete mucus that gives the animal its characteristic sliminess, an adaptation that reduces drag during swimming. In common with sharks, bony fishes have a lateral line system clearly evident as a row of tiny pits in the skin on either side of the body (Figure 6.10).

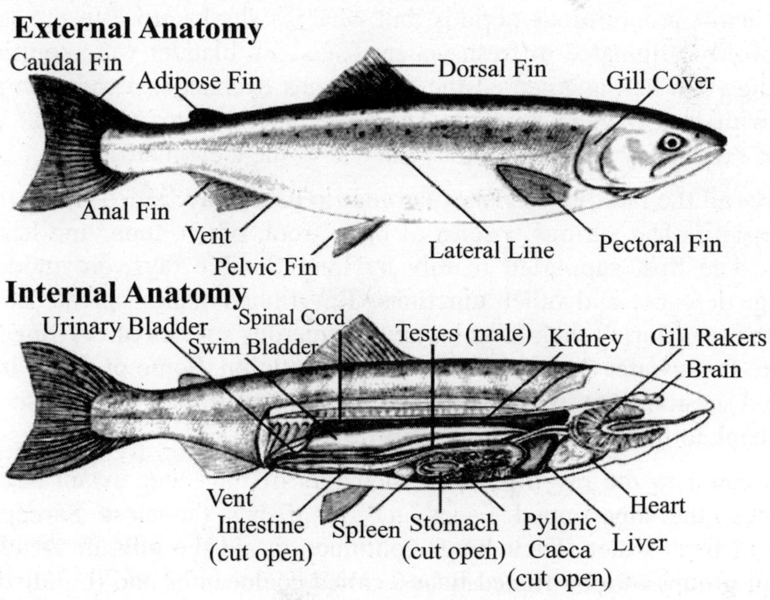

Figure 6.10. Internal and external anatomy of trout, a representative of bony fish.

Bony fishes breathe by drawing water over four or five pairs of gills located in chambers covered by a protective flap called the operculum. Water is drawn into the mouth, through the pharynx, and out between the gills by movement of the operculum and contraction of muscles surrounding the gill chambers. This process enables a bony fish to breathe while stationary.

Another adaptation of most bony fishes not found in sharks is the swim bladder, an air sac that helps control the buoyancy of the fish. The transfer of gases between the swim bladder and the blood varies the inflammation of the bladder and adjusts the density of the fish. Thus, many bony fishes, in contrast to most sharks, can conserve energy by remaining almost motionless. Bony fishes are generally good swimmers, their flexible fins better for steering and propulsion than the stiffer fins of sharks. The fastest bony fishes, which can swim in short bursts of up to 80 km/hr, have the same basic body shape as a shark. In fact, this body shape, termed fusiform (tapering on both ends), is common to all fast fishes and aquatic mammals such as seals and whales. Water is about a thousand times dense than air, and thus the slightest bump that causes drag is even more impeding to a fish than to a bird. Regardless of their different origins, we should expect speedy fishes and marine mammals to have similar streamlined shapes, because the laws of hydrodynamics are universal. This is an example of convergent evolution.

Details about the reproduction of bony fishes vary extensively. Most species are oviparous, reproducing by external fertilization after the female sheds large numbers of small eggs. However, internal fertilization and birthing characterize other species.

Both cartilaginous and bony fishes diversified extensively during the Devonian and Carboniferous periods, but whereas sharks arose in the sea, bony fishes probably originated in fresh water. The swim bladder was modified from simple lungs that had augmented the gills in gas exchange, perhaps in stagnant swamps with low oxygen content. The two major groups (subclass) of bony fishes that exist today had diverged by the end of the Devonian.

Nearly all the families of fishes familiar to us are ray finned fishes (subclass Actinopterygii). The various species of bass, trout, perch, tuna, and herring are examples. The fins, supported mainly by long flexible rays, are modified for swimming, defence, and other functions. Ray-finned fishes spread from fresh water to the seas during their long history. Numerous species of ray-finned fishes returned to fresh water at some point in their evolution. Some of these, including salmon and sea-run trout, replay their evolutionary round-trip from fresh water to seawater back to fresh water during their life cycle.

In contrast to the ray-fins, most members of the other extant subclass of bony fishes, the lobe-finned fishes and lungfishes (subclass Sarcopterygii), remained in fresh water. Their lungs continued to aid the gills in breathing air. Two major groups of lobe-finned fishes, called coelocanths and rhipidistians, are characterized by muscular pectoral and pelvic fins supported by extensions of the bony skeleton. Many lobefins were large, apparently bottom-dwellers that may have used their paired, muscular fins as aids to walking on the substrate under water. Some may also have been able to waddle occasionally on land.

Lobe-finned fishes are represented today by only one known species, a coelacanth named *Latimeria chalumnae* (Figure 6.11). Although most Devonian lobe-fins were probably freshwater animals with lungs, *Latimeria* belongs to a lineage that entered the seas at some point in its evolution.

Figure 6.11. *Latimeria chalumnae*, the only extant lobe-finned fish.

Three genera of lungfishes live today in the Southern Hemisphere. They generally inhabit stagnant ponds and swamps, surfacing to gulp air into lungs connected to the pharynx of the digestive tract. When ponds shrink during the dry season, lungfishes can burrow in the mud and aestivate.

Although most lineages of lungfishes and lobe-fins died out, these groups were dominant predators in shallow freshwater habitats during the Devonian period. They were also of great importance in vertebrate evolution; the ancestor of amphibians was probably one of these Devonian bony fishes.

6.5 SUPERCLASS GNATHOSTOMATA II: THE TETRAPODS

6.5.1 Amphibians are the oldest class of tetrapods

The first vertebrates on land were members of class Amphibia. Today the class is represented by a total of about 4000 species of frogs, salamanders, and caecilians (limbless creatures that burrow in tropical forest and freshwater `lakes).

Early amphibians

As is the case today, parts of the Devonian world were subject to cycles of drought, followed by heavy rainfall and then drought again. In such areas, some lobe-finned fishes with lungs may have prevailed. The skeletal structure of their fins suggests that these paired appendages could have assisted in movement on land (Figure 6.12). Some of the fossil lobe – fins, including a rhipidistian named *Eusthenopteron*, exhibited many other anatomical similarities to the earliest amphibians. A long-standing hypothesis is that Devonian lobe-finned fishes were closely related to the ancestor of tetrapods. However, some molecular evidence suggests that lungfishes are phylogenetically closer to amphibians than are the lobe-fins.

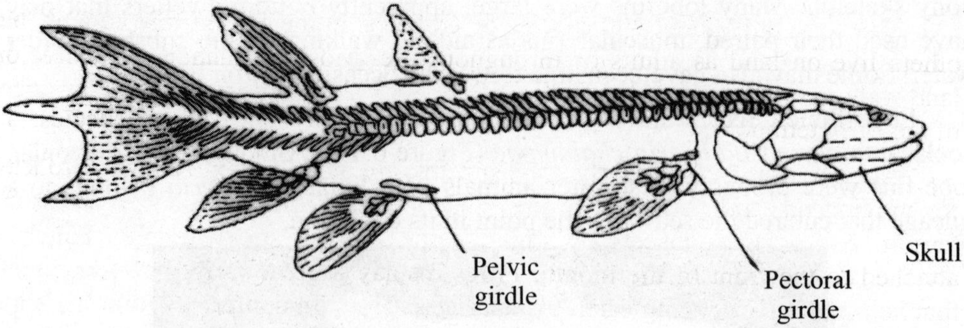

Figure 6.12. The origin of tetrapods. The ancestors of tetrapods may have resembled a lobe-finned fish, *Eusthenopteron.* As indicated in this diagram, muscular fins with extension of the skeleton, the pelvic and pectoral girdle would have provided some support for the animal on land.

The oldest amphibian fossils date back to late Devonian times, about 365 million years ago. The first amphibians may have been primarily aquatic, occasionally venturing on land to escape carnivorous fishes or to exploit a cornucopia of food (insects and other invertebrates) that preceded them onto land. Adaptive radiation of the earliest tetrapods gave rise to a diversity of new forms. Many Carboniferous amphibians superficially resembled reptiles. Some reached 4 m in length. Because amphibians were the only vertebrates on land in late Devonian and early Carboniferous times, the age of amphibians is an appropriate name for the Carboniferous period. Amphibians began to decline during the late Carboniferous. As the Mesozoic era dawned with the Triassic period, about 245 million years ago, most survivors of the amphibian lineage resembled modern species.

Modern Amphibians

There are three extant orders of amphibians (Figure 6.13): Urodela ("tailed ones"—salamanders); Anura ("tail-less ones — frogs, including toads); and Apoda ("legless ones" – caecilians).

Figure 6.13. Amphibian orders. (a) Urodeles (salamander) retain their tails as adults. (b) Anurans, such as frog, lack tails as adults. (c) Apodans, also called caecilians, are legless, mainly burrowing amphibians.

There are only about 400 species of urodeles. Some are entirely aquatic, but others live on land as adults or throughout life. Most salamanders that live on land walk with a side-to-side bending of the body that may resemble the swagger of the early tetrapods.

Anurans, numbering nearly 3500 species, are more specialized than urodeles for moving on land. Adult frogs use their powerful hind legs to hop along the terrain. A frog nabs insects by flicking out its long sticky tongue, which is attached to the front of the mouth. Frogs display a great variety of adaptations that help them avoid being eaten by larger predators. In common with other amphibians, many exhibit color patterns that camouflage. The skin glands of frogs secrete distasteful, or even poisonous, mucus. Many poisonous species have bright coloration that apparently warns predators, who associate the coloration with danger.

Apodans, the caecilians (about 150 species), are legless and nearly blind, and superficially resemble earthworms. Caecilians inhabit tropical areas where most species burrow in moist forest soil; a few South American apodans live in freshwater ponds and streams.

Amphibian means "two lives," a reference to the metamorphosis of many frogs. The tadpole, the larval stage of a frog, is usually an aquatic herbivore with gills, a lateral line system resembling that of fishes, and a long finned tail. The tadpole lacks legs and swims by undulating like its fishlike ancestors. During the metamorphosis that leads to the 'second life', legs develop, and the gills and lateral line system disappear. The young tetrapod with air-breathing lungs, a pair of external eardrums, and a digestive system adapted to a carnivorous diet crawls onto shore and begins life as a terrestrial hunter. In spite of the name amphibian, however, many frogs do not go through the aquatic tadpole stage, and many amphibians do not live a dualistic—aquatic and terrestrial— life. There are some strictly aquatic and strictly terrestrial frogs, salamanders, and caecilians. Moreover, salamander and caecilian larvae look much like adults, and typically both the larvae and the adults are carnivorous. Paedogenesis is common among some groups of salamanders; the mudpuppy (*Necturus*), for instance, retains gills and other larval features when sexually mature.

Most amphibians maintain close ties with water and are most abundant in damp habitats such as swamps and rain forests. Even those frogs that are adapted to drier habitats spend much of their time in burrows or under moist leaves, where the humidity is high. Most amphibians rely heavily on their moist skin to carry out gas exchange with the environment. Some terrestrial species lack lungs and breathe exclusively through their skin and oral cavity.

Amphibian eggs lack a shell and dehydrate quickly in dry air. Fertilization is external in most species, with the male grasping the female and spilling his sperm over the eggs as the female sheds them. Amphibians generally lay their eggs in ponds or swamps or at least in moist environments. Some species lay vast numbers of eggs in temporary pools, and mortality is high. In contrast, some species display various types of parental care and lay relatively few eggs. Depending on the species, either males or females may incubate eggs on their back, in the mouth, or even in the stomach. Certain tropical tree frogs stir their egg masses into moist foamy nests that resist drying. There are also some ovoviviparous and viviparous species that retain the eggs in the female reproductive tract, where embryos can develop without drying out.

Many amphibians exhibit complex and diverse social behavior, especially during their breeding seasons. Frogs are usually quiet creatures, but many species fill the air with their mating calls during the breeding season. Males may vocalize to defend breeding territory or to attract females. In some terrestrial species, migrations to specific breeding sites may involve vocal communication, celestial navigation, or chemical signaling.

6.5.2 Evolution of the amniotic egg expanded the success of vertebrates on land

The evolution of reptiles from an amphibian ancestor involved many adaptations for terrestrial living. The amniotic egg, a reproductive adaptation that enabled terrestrial vertebrates to complete their life cycles on land and sever their last ties with their aquatic origins, was a particularly important breakthrough. In contrast to the shell-less eggs of amphibians, amniotic eggs (of birds, most reptiles, and some mammals) have a shell that retains water and can therefore be laid in a dry place. Most mammals have dispensed with the shell; instead, the embryo implants in the wall of the uterus and obtains its nutrients from the mother.

Reptiles, birds, and mammals all have specialized membranes within the amniote egg, called extra-embryonic membranes (the amnion, chorion, allantois and yolk sac) because they are not part of the body of the developing animal (Figure 6.14). The amnion protects the embryo in a fluid filled cavity that prevents dehydration and cushions mechanical shocks. The yolk sac expands over the yolk, a stockpile of nutrients stored in egg. Blood vessels in the yolk sac membrane transport nutrients from the yolk into the embryo. Other nutrients are stored in the albumin (the egg white). The allantois functions as disposal sac for certain metabolic wastes produced by the embryo. The membrane of the allantois also functions with the chorion as a respiratory organ that exchanges gases between the embryo and the surrounding air. O_2 and CO_2 diffuse freely across the eggs shell. However, the shell is waterproof, an adaptation that prevents the embryo and the fluid compartments around it from drying out on land. These extraembryonic membranes develop from tissue layers that grow out from the embryo. The amniotic egg is named after one of these membranes, the amnion, which encloses a compartment of amniotic fluid that bathes the embryo and acts as a hydraulic shock absorber. Together, reptiles, birds, and mammals make up a monophyletic group, the amniotes.

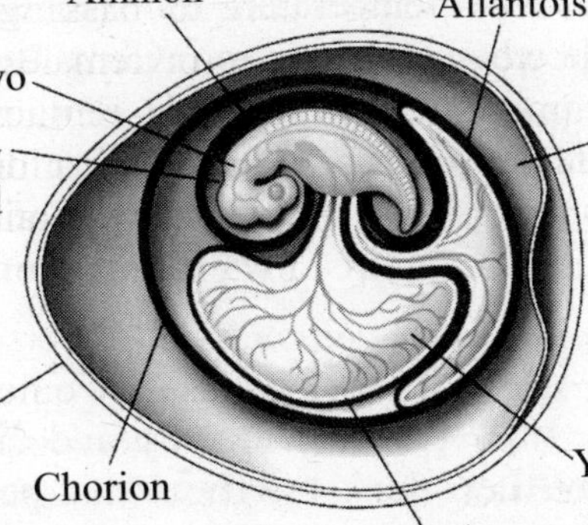

Figure 6.14. The amniotic egg. The embryos of reptiles, birds, and mammals form four extraembryonic membranes; the amnion, chorion, allantois, and yolk sac.

6.5.3 Reptilian evolution

Class reptilian, a diverse group with many extinct lineages, is represented today by about 7000 species, mainly lizards, snakes, turtles, and crocodilians. This grouping is traditional and based on the apparent similarity of all these tetrapods. Cladistic analysis indicates, however, that grouping all these vertebrates in a class that does not include birds is inconsistent with phylogeny. Birds appear to be more closely related to crocodiles than are turtles. In fact, Class Reptilia cannot be defined except by the absence of features that distinguish birds (flight feathers) and mammals (hair and mammary glands). Despite these taxonomic problems, studying reptiles (in the traditional sense) provides a context for understanding all amniotes. Hence, we follow tradition in this case and recognize the familiar classes Reptilia, Aves, and Mammalia.

Reptilian Characteristics

Reptiles have several adaptations for terrestrial living not generally found in amphibians. Scales containing the protein keratin waterproof the skin of a reptile, helping prevent dehydration in dry air. Keratinized skin is the vertebrate analogue of the chitinized cuticle of insects and the waxy cuticle of land plants. Because they cannot breathe through their dry skin, most reptiles obtain all their oxygen with lungs. Many turtles also use the moist surfaces of their cloaca for gas exchange.

Most reptiles lay shelled amniotic eggs on land. Fertilization in reptiles must occur internally, before the shell is secreted as the egg passes through the reproductive tract of the female. Some species of snakes and lizards are viviparous, their extra embryonic membranes forming a placenta that enables the embryo to obtain nutrients from its mother.

Reptiles are sometimes labeled "cold-blooded" animals because they do not use their metabolism extensively to control body temperature. But reptiles do regulate body temperature by using behavioral adaptations. For example, many lizards regulate their internal temperature by basking in the sun when the air is cool and seeking shade when the air is too warm. Because they absorb external heat rather than generating much of their own, reptiles are said to be ectotherms, a term more appropriate than cold-blooded. By heating directly with solar energy rather than through the metabolic breakdown of food, a reptile can survive on less than 10% of the calories required by a mammal of equivalent size.

The Age of Reptiles

Reptiles were far more widespread, numerous, and diverse during the Mesozoic era than they are today. Let's look briefly at this period, known as the age of reptiles.

The Origin and Early Evolutionary Radiation of Reptiles

The oldest reptilian fossils are found in rocks from the late Carboniferous period; they are about 300 million years old. Their ancestor was among the Devonian amphibians. In two great waves of adaptive radiation, reptiles became the dominant terrestrial vertebrates in a dynasty that lasted more than 200 million years.

The first major reptilian radiation occurred by the dawn of the Permian, the last period of the Paleozoic era, giving rise to two main evolutionary branches (Figure 6.15):

1. **The synapsids**. This branch included a diversity of mammal-like reptiles called therapsids, including the probable ancestor of mammals.
2. **The sauropsids**. This branch gave rise to all modern amniotes except the mammals. Relatively early in their history, the sauropsids split into two sub-branches:
   a. *The anapsids*. Turtles are the only survivors of this reptilian group.
   b. *The diapsids*. Lizards, snakes, and crocodilians are the extant diapsids classified as reptiles. The dinosaurs and some other groups of extinct reptiles were also diapsids. Cladistic analysis provides strong evidence that birds are the closest living relatives of the extinct dinosaurs.

Dinosaurs and Pterosaurs

The second great reptilian radiation was under way by late Triassic times (a little more than 200 million years ago) and was marked mainly by the origin and diversification of two groups of reptiles: the dinosaurs, which lived on land, and the pterosaurs, or flying reptiles. These groups were the dominant vertebrates on earth for millions of years. Pterosaurs had wings formed from a membrane of skin stretched from the body wall, along the forelimb, to the tip of an elongated finger. Stiff fibers provided support for the skin of the wing. Dinosaurs, an extremely diverse group varying in body shape, size, and habitat, included the largest animals ever to inhabit land. Fossils of gigantic dinosaurs that are 45 m long have recently been unearthed in New Mexico and Utah.

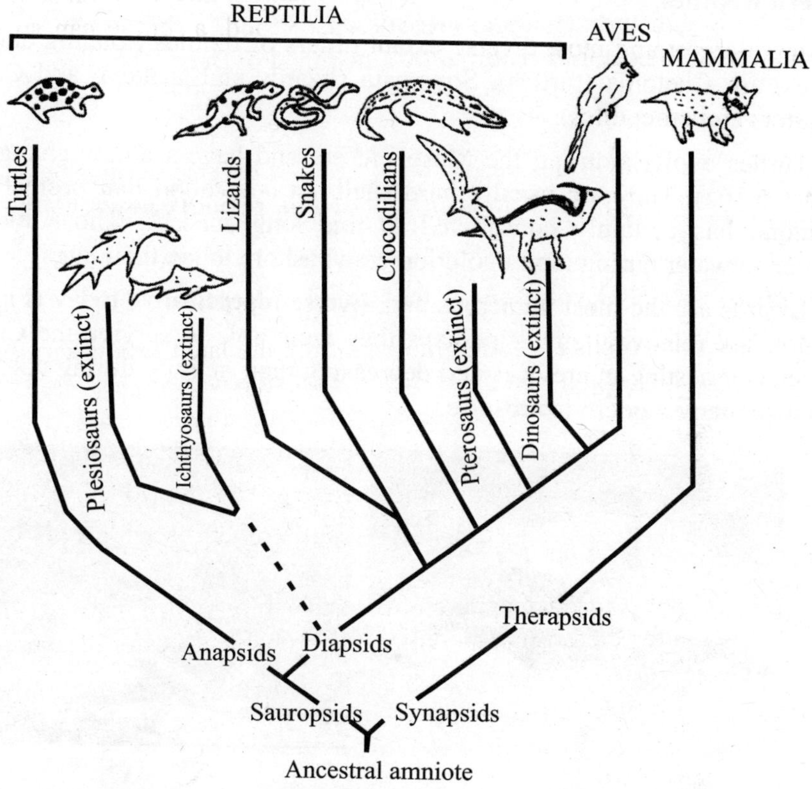

Figure 6.15. Phylogeny of amniotes.

Contradicting the long-standing view that dinosaurs were slow, sluggish creatures, there is increasing evidence that many dinosaurs were agile, fast-moving, and, in some species, social. Paleontologists have also discovered signs of parental care among dinosaurs. There is continuing debate about whether dinosaurs were endothermic, capable of keeping the body warm through metabolism. Some anatomical evidence supports this hypothesis, but most experts are skeptical. The Mesozoic climate was relatively warm and consistent, and behavioral adaptations such as basking may have been sufficient for maintaining a suitable body temperature, especially for the land-dwelling dinosaurs. Also, large dinosaurs had low surface-to-volume ratios that reduced the effects of daily fluctuations in air temperature on the internal temperature of the animal.

The Cretaceous Crisis

During the Cretaceous period, the last period of the Mesozoic era, the climate became cooler and more variable. This was a period of mass extinctions, and except for a few dinosaurs that survived into the early Cenozoic period, all of these reptiles were gone by the end of the Cretaceous period, about 65 million years ago (Cretaceous mass extinction has been discussed in Chapter III).

Modern Reptiles

The three largest and most diverse extant orders of reptiles (totaling about 6500 species) are Chelonia (turtles), Squamata (lizards and snakes), and Crocodilia (alligators and crocodiles).

Turtles evolved during the Mesozoic era and have scarcely changed since (Figure 6.16a). They are usually hard shell, an adaptation that protects against predators, has certainly contributed to this long success. Those turtles that returned to water during their evolution crawl ashore to lay their eggs.

Lizards are the most numerous and diverse reptiles alive today (Figure 6.16 b). Most are relatively small; perhaps they were able to survive the Cretaceous "crunch" by nesting in crevices and decreasing their activity during cold periods, a practice many modern lizards use.

(a) (b) (c) (d)

Figure 6.16. Modern reptiles (a) Turtles (b) Lizards (c) Snakes and (d) Crocodiles.

Snakes are apparently descendants of lizards that adopted a burrowing lifestyle (Figure 6.16c). Today, most snakes live above the ground, but they have retained the limbless condition. Vestigial pelvic and limb bones in primitive snakes such as boas, however, are evidence that snakes evolved from reptiles with legs.

Snakes are carnivorous, and a number of adaptations aid them in hunting prey. They have acute chemical sensors, and though they lack eardrums, snakes are sensitive to ground vibrations, which help them detect the movements of prey. Heat-detecting organs between the eyes and nostrils of pit vipers, including rattlesnakes, are sensitive to minute temperature changes, enabling these night hunters to locate warm animals. Poisonous snakes inject their toxin through a pair of sharp hollow or grooved teeth. The flicking tongue is not poisonous but helps fan odors toward olfactory organs on the roof of the mouth. Loosely articulated jaws enable most snakes to swallow prey larger than the diameter of the snake itself.

Crocodiles and alligators (crocodilians) are among the largest living reptiles (some turtles are heavier) (Figure 6.16 d). They spend most of their time in water, breathing air through their upturned nostrils. Crocodilians are confined to the warm regions of Africa, China, Indonesia, India, Australia, South America, and the southeastern United States, where alligators have made a strong comeback after spending years on the endangered species list. Among the modern animals traditionally classified as reptiles, crocodilians are the most closely related to the dinosaurs. However, the modern animals that seem to share the most recent ancestor with the dinosaurs are the birds.

6.5.4 Evolution of Birds

Class Aves (birds) evolved during the great reptilian radiation of the Mesozoic era. Amniotic eggs and scales on the legs are just two of the reptilian features we see in birds. But modern birds look quite different from modern reptiles because of their feathers and other distinctive flight equipment.

Characteristics of Birds

Almost every part of a typical bird's anatomy is modified in some way that enhances flight. The bones have an internal structure that is honeycombed, making them strong but light (Figure 6.17). The skeleton of a frigate bird, for instance, has a wingspan of more than 2 m but weighs only about 113 g. Another adaptation reducing the weight of birds is the absence of some organs. Females, for instance, have only one ovary. Also, modern birds are toothless, an adaptation that trims the weight of the head. Food is not chewed in the mouth but ground in the gizzard, a digestive organ near the stomach (Crocodiles also have gizzards, as did some dinosaurs). The bird's beak, made of keratin, has proven to be very adaptable during avian evolution, taking on a great variety of shapes suitable for different diets.

Flying requires a great expenditure of energy from an active metabolism. Birds are endothermic; they use their own metabolic heat to maintain a warm, constant body temperature. Feathers and in some species a layer of fat provides insulation that enables birds to retain their metabolically generated heat. An efficient respiratory system and a circulatory system with a four-chambered heart keep tissues well supplied with oxygen and nutrients, supporting a high rate of metabolism. The efficient lungs have tiny tubes leading to and from elastic air sacs that help dissipate heat and reduce the density of the body. For safe flight, senses, especially vision, must be acute. Birds have excellent eyes, perhaps the best of all vertebrates. The visual areas of the brain are well developed, as are the motor areas; flight also requires excellent coordination.

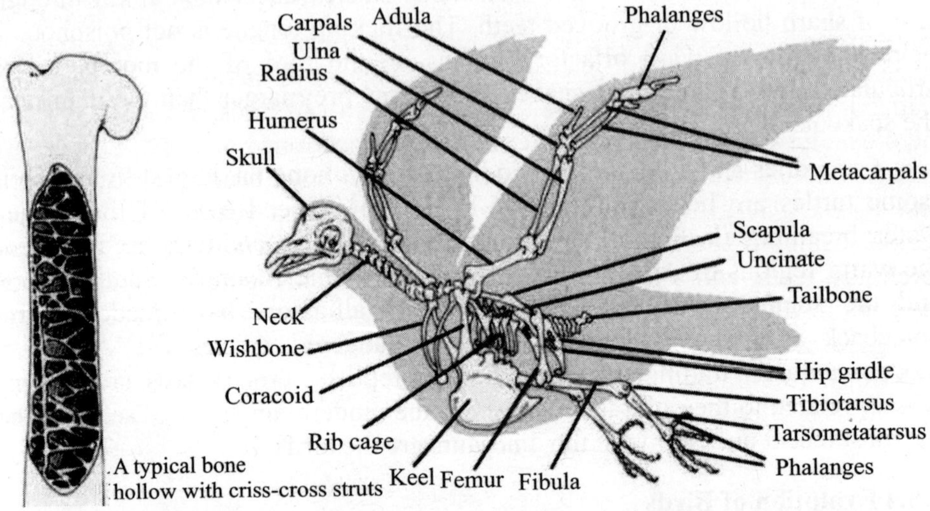

Figure 6.17. The avian skeleton system. The avian skeleton system showing the honeycombed anatomy of the bones on the left is a flight adaptation.

With brains proportionately larger than those of reptiles and amphibians, birds generally display very complex behavior. Avian behavior is particularly intricate during breeding season, when birds engage in elaborate rituals of courtship. Since eggs are shelled when laid, fertilization must be internal. Copulation involves contact between the mates' vents, the openings to their cloacas. After eggs are laid, the avian embryo must be kept warm through brooding by the mother, father, or both, depending on the species.

A bird's most obvious adaptation for flight is its wings. Bird wings are airfoils that illustrate the same principles of aerodynamics as the wings of an airplane. Providing power for flight, birds flap their wings by contractions of large pectoral (breast) muscles anchored to a keel on the sternum (breastbone). Some birds, such as eagles and hawks, have wings adapted for soaring on air currents and flap their wings only occasionally; other birds, including humming birds, must flap continuously to stay aloft. In either case, it is the shape and arrangement of the feathers that form the wing into an airfoil.

In being both extremely light and strong, feathers are among the most remarkable of vertebrate adaptations. Feathers are made of keratin, the same protein that forms our hair and fingernails and the scales of reptiles. Feathers may have functioned first as insulation during the evolution of endothermy, only later being co-opted as flight equipment. Besides providing support and wing shape, feathers can be manipulated to control air movements around the wing.

The evolution of flight required radical alteration in body form, but flight provides many benefits. It enhances hunting and scavenging; many birds exploit flying insects, an abundant, highly nutritious food resource. Flight also provides ready escape from earthbound predators and enables some birds to migrate great distances to utilize different food resources and seasonal breeding areas.

The Origin of Birds

The presence of feathers may be enough to classify an animal as a bird, although some recently discovered dinosaur fossils appear to have short, down like feathers. Certainly, if we want to trace the ancestry of birds, we must search for the oldest fossils with feathers that could have functioned in flight. Fossils of an ancient bird named *Archaeopteryx lithographica* have been found in Bavarian limestone in Germany, dating back some 150 million years, into the Jurassic period. Unlike modern birds, but similar to reptiles, *Archaeopteryx* had clawed forelimbs, teeth, and a long tail containing vertebrae (Figure 6.18). Indeed, if it were not for the preservation of its feathers, *Archaeopteryx* would be regarded as a member of a diverse group of small, carnivorous, bipedal dinosaurs called theropods. In common with birds, many dinosaurs, including some theropods, built nests and cared for their young.

Figure 6.18. *Archeopteryx*, a Jurassic bird-reptile.

Archaeopteryx is not considered the ancestor of modern birds, and paleontologists place it on a side branch of the avian lineage. Nonetheless, *Archaeopteryx* probably was derived from ancestral forms that also gave rise to modern birds. Its skeletal anatomy indicates that it was a weak flyer, perhaps mainly a tree-dwelling slider. A combination of gliding downward and jumping into the air from the ground may have been the earliest mode of flying in the bird lineage.

Extensive cladistic analyses have convinced many researchers that birds arose from theropod dinosaurs. Heated debate continues, however, with some biologists arguing that the bird lineage diverged from a group of early Mesozoic reptiles, called thecodonts, that also gave rise to the dinosaurs.

Modern Birds

There are about 8600 extant species of birds classified in about 28 orders. Flight is typical of birds, but there are several flightless species, including the ostrich, and kiwi. Flightless birds are collectively called **ratites** because their breastbone lacks a keel and the large breast muscles that attach to it in flying birds.

In contrast to ratites, other birds are called **carinates** because they have a carina, or sternal keel, supporting their large breast muscles. These muscles provide flight power in flying birds. The demands of flight have rendered the general body form of carinate birds similar to one another, yet experienced bird watchers can distinguish many species by their body profile. Carinate birds also exhibit great variety in feather colors, beak and foot shape, behavior, and flying style. Among the most unusual carinate birds are the penguins, which do not fly, but use their powerful breast muscles in swimming. Nearly 60% of the living bird species belong in one carinate order called the **passeriformes**, or perching birds. These are the familiar jays, swallows, sparrows, warblers, and many others.

6.5.5 Mammalian evolution

The extinction of the dinosaurs and the fragmentation of continents that occurred at the close of the Mesozoic era opened many adaptive zones, and mammals underwent an extensive adaptive radiation. There are about 4500 species of mammals on earth today. As mammals ourselves, we naturally have a special interest in this class of vertebrates. Let's examine some of the features we share with all other mammals.

Mammalian Characteristics

Vertebrates of **Class Mammalia** have hair, a characteristic as diagnostic as the flight feathers of birds. Hair, like feathers, is made of keratin, although zoologists are uncertain about its evolutionary origin. Most mammals have an active metabolism and are endothermic. Efficient respiratory and circulatory systems (including a four-chambered heart) support a high metabolic rate. A sheet of muscle called the diaphragm helps ventilate the lungs. Hair and a layer of fat under the skin help the body retain metabolic heat.

Mammary glands that produce milk are as distinctively mammalian as hair. All mammalian mothers nourish their babies with milk, a balanced diet rich in fats, sugars, proteins, minerals, and vitamins.

Most mammals are born rather than hatched. Fertilization is internal, and the embryo develops inside the uterus of the female's reproductive tract. In eutherian (placental) mammals and marsupials, the lining of the mother's uterus and extraembryonic membranes arising from the embryo collectively form a placenta, where nutrients diffuse into the embryo's blood. Eutherians are commonly called placental mammals because their placentas are more complex and provide a more intimate and long-lasting association between the mother and her developing young.

Mammals generally have larger brains than other vertebrates of equivalent size, and many species are capable learners. The relatively long duration of parental care extends the time for offspring to learn important survival skills by observing their parents.

Differentiation of teeth is another important mammalian trait. Whereas the teeth of reptiles are generally conical and uniform in size, the teeth of mammals come in a variety of sizes and shapes adapted for chewing many kinds of foods. Our own dentition, for example, includes teeth modified for shearing (incisors and canine teeth) and for crushing and grinding (premolars and molars). Another part of the feeding apparatus, the jaw, was also remodeled during the evolution of mammals from reptiles, and two of the jawbones were incorporated into the mammalian inner ear.

The Evolution of Mammals

Mammals evolved from reptilian stock even earlier than birds. The oldest fossils believed to be mammalian date back 220 million years, into the Triassic period. The ancestor of mammals was among the therapsids, part of the synapsid branch of reptilian phylogeny. The therapsids disappeared during the reign of the dinosaurs, but their mammalian offshoots coexisted with the dinosaurs throughout the Mesozoic era. Most Mesozoic mammals were very small— about the size of shrews—and most probably ate insects. A variety of evidence, such as the size of the eye sockets, suggests that these tiny mammals were nocturnal.

As the Cenozoic era dawned in the wake of the Cretaceous extinctions, mammals were in the midst of a great adaptive radiation. That diversity is represented today in the three major groups: monotremes (egg-laying mammals), marsupials (mammals with pouches), and eutherian (placental) mammals.

Monotremes

Monotremes, the platypuses and the echidnas (spiny anteaters)—are the only living mammals that lay eggs (Figure 6.19a). The egg, which is reptilian in structure and development, contains enough yolk to nourish the developing embryo. Monotremes have hair and produce milk, two of the most important trademarks of Mammalia. On the belly of a monotreme mother are specialized glands that secrete milk. After hatching, the baby sucks the milk from the fur of the mother, who has no nipples. The mixture of ancestral reptilian characters and derived characters of mammals suggests that monotremes descended from a very early branch in the mammalian genealogy. Today, monotremes are found only in Australia and New Guinea.

(a)

(b)

Figure 6.19. Australian monotremes and marsupials, (a) Monotreme: A short-beaked Echidna. (b) The young of marsupials (Kangaroo) within a maternal pouch.

Marsupials

Opossums, kangaroos, bandicoots, and koalas are examples of marsupial mammals. A marsupial is born very early in its development and completes its embryonic development while nursing (Figure 6.19b). In most species, the nursing young are held within a maternal pouch called a marsupium. A red kangaroo, for instance, is about the size of a honeybee at its birth, just 33 days after fertilization. Its hind legs are merely buds, but the forelimbs are strong enough for the offspring to crawl from the exit of the reproductive tract to the mother's pouch, a journey lasting a few minutes.

In Australia, marsupials have radiated and filled niches occupied by eutherian (placental) mammals in other parts of the world. Convergent evolution has resulted in a diversity of marsupials that resemble their eutherian counterparts occupying similar ecological roles.

The opossums of North and South America are the most diverse of only three families of extant marsupials outside the Australian region, though South America had a diverse marsupial fauna throughout the Tertiary period. The distribution of modern and fossil marsupials begins to make sense in the context of plate tectonics and continental drift. According to recent fossil evidence, marsupials probably originated in what is now North America, spreading southward when the land masses were still joined. After the breakup of Pangaea, South America and Australia became island continents, and their marsupials diversified in isolation from the placental mammals that began an adaptive radiation on the northern continents. Australia has not been in contact with another continent since early in the Cenozoic era, about 65 million years ago. The South American fauna, meanwhile, has not remained cloistered. Placental mammals reached South America throughout the Cenozoic. The most important migrations occurred about 12 million years ago and then again about 3 million years ago, when North and South America joined at the Panamanian isthmus. Extensive two-way traffic of animals took place over the land bridge. The biogeography of mammals is another example of the interplay between biological and geological evolution.

Eutherian (Placental) Mammals

Compared to marsupials, eutherian mammals have a longer period of pregnancy. Young eutherians complete their embryonic development within the uterus, joined to the mother by the placenta.

Adaptive radiation during late Cretaceous and early Tertiary periods (about 70 to 45 million years ago) produced the orders of eutherian mammals we recognize today. The phylogenetic relationship between marsupials and eutherian mammals is somewhat obscure, but scientists believe they are more closely related to each other than either is to the monotremes. Fossil evidence indicates that eutherians and marsupials may have diverged from a common ancestor about 80 to 100 million years ago.

The evolutionary relationships among the many orders of eutherian mammals are, for the most part, also unsettled. Most mammalogists now favour a tentative genealogy that recognizes at least four main evolutionary lines of eutherian (placental) mammals.

One branch of eutherian mammals consists of the order Insectivora (shrews), which resemble early mammals, and the order Chiroptera (bats). The bats, whose forelimbs are modified as wings, probably evolved from insectivores that fed on flying insects. In addition to the insect-eaters, some bat species feed on small vertebrates or fruit; three species of vampire bats found in Central and South America bite mammals and lap up their blood. Most bats are nocturnal.

The second branch started with a lineage of medium-sized herbivores that underwent a spectacular adaptive radiation during the Tertiary period, eventually giving rise to such modern orders as Lagomorpha (rabbits and their relatives); Perissodactyla (odd-toed ungulates, including horses and rhinoceroses; ungulates walk on the tips of toes); Artiodactyla (even-toed ungulates, including deer and swine); Sirenia (sea cows); Proboscidea (elephants); and Cetacea (porpoises and whales).

A third branch in the evolution of eutherian mammals produced the order Carnivora, which includes cats, dogs, raccoons, skunks, and the pinnipeds (seals, sea lions, and walruses). Carnivorans probably first appeared during the early Cenozoic era. Seals and their relatives apparently evolved from middle Cenozoic carnivorans that became adapted for swimming.

A fourth and very extensive adaptive radiation of eutherian mammals produced the primate-rodent complex. Order Rodentia includes rats, mice, squirrels, and beavers. With about 1770 species, this is by far the largest order of mammals. The name Rodentia, from the Latin word for "gnawing' refers to one of the distinctive features of the group: both the upper and the lower jaw have a pair of large front teeth (incisors) that resist heavy wear by growing continually. Order Primates includes monkeys, apes and humans, which we discuss next.

6.6 PRIMATES AND THE EVOLUTION OF *HOMO SAPIENS*

Having tracked the vertebrate genealogy to the mammalian orders, we now begin to trace our own specific ancestry. Order Primates includes Homo sapiens and its closest kin. We are primates. To learn what that means, we must trace our ancestry back to the trees, where some of our most treasured traits had their beginnings as adaptations to an arboreal (tree-dwelling) existence.

6.6.1 Primate evolution provides a context for understanding human origins

Dental structure suggests that the first primates descended from insectivores in the Cretaceous period. These early primates were probably small arboreal (tree-dwelling) mammals. Thus, during the Mesozoic era, our order was defined by characteristics that had been shaped, through natural selection, by the demands of living in the trees. For example, primates have limber shoulder joints, which make it possible to brachiate (swing from one hold to another). The dexterous hands of primates can hang on to branches and manipulate food. Claws have been replaced by nails in many species, and the fingers are very sensitive. The eyes of primates are close together on the front of the face. The overlapping fields of vision of the two eyes enhance depth perception, an obvious advantage when brachiating. Excellent eye-hand coordination is also important for arboreal maneuvering. Parental care is essential for young animals in the trees. Mammals devote more energy to caring for their young than most other vertebrates, and primates are among the most attentive parents of all mammals. Most primates have single births and nurture their offspring for a long time. Though humans do not live in trees, we retain in modified form many of the traits that evolved there.

Primate Classification

There are presently about 185 primate species, which primatologists usually classify (Table 6.1) into two suborders, **Prosimii** and **Anthropoidea**. The prosimians ("premonkeys") probably resemble early arboreal primates and generally retain more early mammalian features (ex., claws, long snout, and lateral-facing eyes) than do the anthropoid primates (Figure 6.20). The lemurs of Madagascar and the lorises, pottos, and tarsiers that live in tropical Africa and southern Asia are examples of prosimians. With exception of Medagascar, prosimians in other localities have small bodies and are nocturnal.

The anthropoids, which include monkeys, apes, and humans, are mostly larger than prosimians and generally diurnal rather than nocturnal. Compared to prosimians, anthropoids also have more of the primate features such as a shortened face, forward looking eyes, and a larger and more complex brain.

The evolution of anthropoids from prosimians is a topic of active debate. Paleontologists divide prosimian fossils into two groups: one ancestral to tarsiers, the other ancestral to lemurs, lorises, and pottos. This split occurred at least 50 million years ago. Until recently, the choice between these two branches as the lineage that also gave rise to anthropoids. However, some recently discovered fossils in Asia and Africa may be more similar to anthropoids than are either of the other two groups of prosimian fossils. The age of these fossils is still in dispute, but their discoverers date them back at least 50 million years. These discoveries raise the possibility that the prosimians had already split into at least three lineages by 50 million years ago, with the anthropoid ancestry already differentiated from the two branches leading to modern prosimians. Thus, the current debate about anthropoid origins can be phrased this way: Are anthropoids more closely related to tarsiers, to lemurs and lorises, or to some third group of prosimians lacking extant representatives? More fossils will help answer this question.

Table 6.1. Classification of existing subgroups in the order Primates, with common names for some members in each group

Classification
Suborder: Prosimii
Superfamily: Lemuroidea (lemurs)
Superfamily: Lorisoidea (lorises, galagos)
Superfamily: Tarsioidea (tarsiers)
Suborder: Anthropoidea
Infraorder: Platyrrhini (New World)
Superfamily: Ceboidea
Family: Callitrichidae (marmosets, tamarins)
Family: Cebidae (capuchins, spider monkeys)
Infraorder: Catarrhini (Old World)
Superfamily: Cercopithecoidea
Family: Cercopithecidae (macaques, baboons)
Family: Colobidae (langurs)
Superfamily: Hominoidea
Family: Hylobatidae (gibbons)
Family: Pongidae (orangutans, gorillas, chimpanzees)
Family: Hominidae (humans)

Fossils of monkey like primates indicate that anthropoids were already established in Africa and Asia by 40 million years ago. Africa and South America had already drifted apart, and it is not clear whether ancestors of New World monkeys reached South America by rafting on logs or other debris from Africa or by migration southward from North America. What is certain is that New World monkeys and Old World monkeys have been evolving along separate pathways for many millions of years (Figure 6.21). All New World monkeys are arboreal, whereas Old World monkeys include ground-dwelling as well as arboreal species. Most monkeys of both groups are diurnal (active during the day) and usually live in bands held together by social behavior.

In addition to monkeys, the anthropoid suborder also includes the four genera of apes, shown in Figure 6.22, Hylobates (gibbons), Pongo (orangutans), Gorilla (gorillas), and Pan (chimpanzees). Modern apes are confined exclusively to tropical regions of the Old World. With the exception of gibbons, modern apes are larger than monkeys, with relatively long arms and short legs and no tails. Although all the apes are capable of brachiation, only gibbons and orangutans are primarily arboreal. Social organization varies among the genera - of apes; gorillas and chimpanzees are highly social. Apes have larger brains proportionate to body size than monkeys, and their behavior is consequently more adaptable.

Figure 6.20. A crowned Madagascar lemur, a prosimian.

(a) New World monkey (spider monkey) (b) Old World monkey (baboon)

Figure 6.21. New world and Old world monkeys. (a) New World monkeys, such as spider monkeys have prehensile tails and nostrils that open to the sides. (b) Old world monkeys, such as baboons lack prehensile tails and their nostrils open downward

(a) Gibbon

(b) Orangutan

(c) Gorilla

(d) Chimpanzee

Figure 6.22. Apes. (a) Gibbons are almost arboreal; they have relatively small size with long arms. (b) Orangutans are large apes; mostly spend their time on trees and lives in the rain forest of Sumatra and Borneo. (c) The gorillas are the largest apes; they are mainly confined to Africa. (d) The chimpanzees live in tropical Africa. Chimpanzees feed and sleep on trees but also spend a great deal of time on the ground; they are intelligent, communicative and social

6.6.2 Humanity: a very young twig on the vertebrate tree

In the continuum of life spanning over 3.5 billion years, humans and apes have shared ancestry for all but the last few million years. Paleoanthropology, the study of human origins and evolution, focuses on this tiny fraction of geological time during which humans and chimpanzees diverged from a common ancestor.

Some Common Misconceptions

Paleoanthropology has a checkered history. Until about 20 years ago, researchers often gave new names to fossil forms that were undoubtedly the same species as fossils found by competing scientists. Elaborate theories have often been based on a few teeth or a fragment of jawbone. Many misconceptions about human evolution generated during the early part of the twentieth century still persist in the minds of much of the general population, long after these myths have been replaced by fossil discoveries.

Let's first dispose of the myth that our ancestors were chimpanzees or any other modern apes. Chimpanzees and humans represent two divergent branches of the anthropoid tree that evolved from a common, less-specialized ancestor.

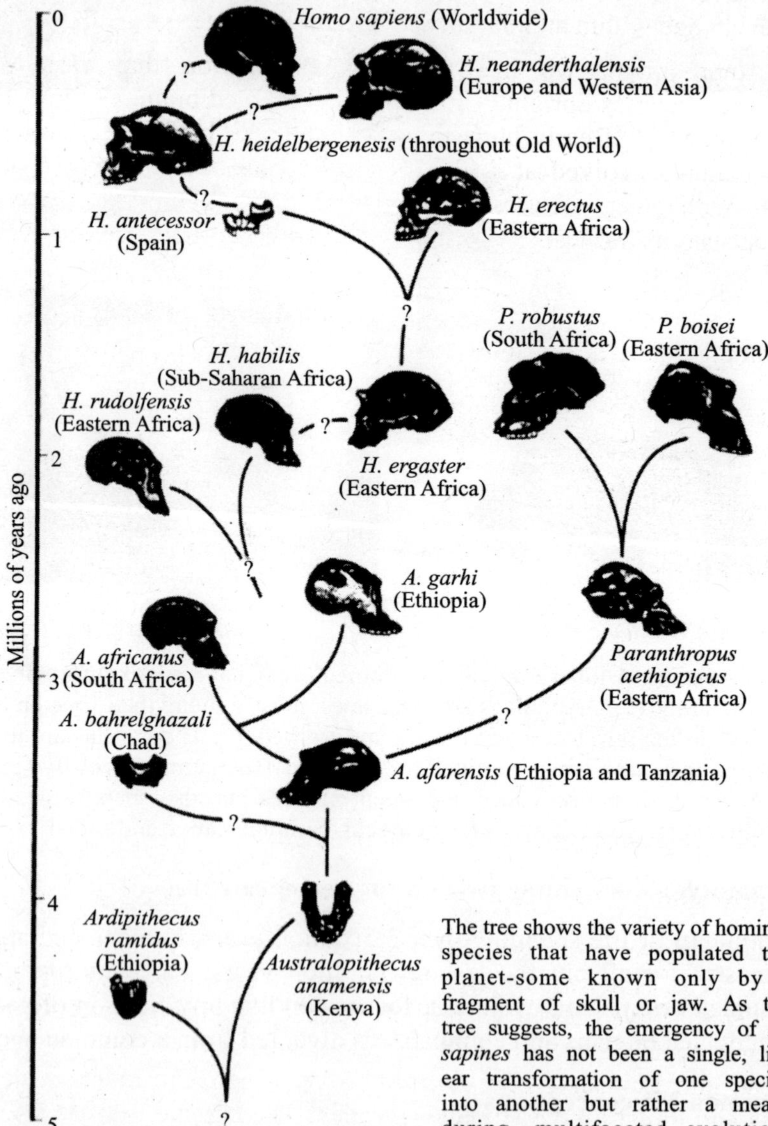

The tree shows the variety of hominid species that have populated the planet-some known only by a fragment of skull or jaw. As the tree suggests, the emergency of *H. sapines* has not been a single, linear transformation of one species into another but rather a meanduring, multifaceted evolution.

Figure 6.23. A time line of some hominid species

Another misconception envisions human evolution as a ladder with a series of steps leading directly from an ancestral anthropoid to *Homo sapiens*. This is often illustrated as a parade of fossil hominids (members of the human family) becoming progressively more modern as they march across the page. If human evolution is a parade, then it is a disorderly one, with many splinter groups having traveled down dead ends. At times in hominid history, several different human species coexisted (Figure 6.23). Human phylogeny is more like a multibranched bush than a ladder, our species being the tip of the only twig that still lives. If the theory of punctuated equilibrium applies to humans, then most change occurred as new hominid species came into existence, and not by phyletic (anagenic) change within an unbranched hominid lineage.

One more myth we must bury is the notion that various human characteristics, such as upright posture and an enlarged brain, evolved in unison. A popular image is of early humans as half-stooped, half-witted cave dwellers. Different features evolved at different rates--a phenomenon known as mosaic evolution—with erect posture, or bipedalism, leading the way. Our pedigree includes ancestors who walked upright but had a brain much less developed than ours.

After dismissing some of the folklore on human evolution, however, we must admit that many questions about our ancestry remain.

Figure 6.24. *Aegyptopithecus*, a cat-sized tree-dweller.

Early Anthropoids

The oldest known fossils of apes have been assigned to the genus *Aegyptopithecus*, (Figure 6.24) the "dawn ape." This anthropoid was a cat-sized tree-dweller that lived about 35 million years ago. During the Miocene epoch, which began about 23 million years ago, descendants of the first apes diversified and spread into Eurasia. About 20 million years ago, the Indian plate collided with Asia and thrust up the Himalayan range. The climate became drier and the forests of what is now Africa and Asia contracted, isolating these two regions of anthropoid evolution. Among the African anthropoids was the common ancestor of humans and chimpanzees. Based on the fossil record and comparisons of DNA between humans and chimpanzees, most anthropologists now agree that humans and apes diverged from this common ancestor only about 6 to 8 million years ago.

6.6.3 Four main classes of change occurred during hominin evolution

Humans are primates, and our ancestors from about 60 (or more) million years ago to about 5-10 million years ago were tree dwellers. The big changes in human evolution may have occurred after our ancestors moved from forests to more savannah-type habitats and becomes land dwellers. In fossils, the big change can be understood in three categories. A fourth category concerns changes in social behaviour, which is less easily studied in fossils.

1. *Brain enlargement*. Modern chimpanzees have brains of about 350-400 cm^3, and our ape ancestors 5 million years ago probably had brains of about the same size. Modern human brains are about 1,350 cm^3 in size. This suggests that there is a progressive increase in the size of brain during the course of human evolution.
2. *Changes to the jaws and teeth*. Chimpanzees, and our ape ancestors, are more prognathic than us, with their jaws sticking out more from their faces. During human evolution, the jaw shrunk back into the face, giving us flat faces. The jaw of our ape ancestors, and chimpanzees has a semicircular shape (if viewed from above or below). In us, the semicircle has been pushed back to a shape more like a rectangle. Our teeth also got smaller, particularly the canine teeth, and our molars have evolved into grinding millstones.
3. *Bipedality*. The evolution of upright locomotion on two legs has resulted in changes through out our bodies. Adaptations for bipedality are particularly clear in the anatomy of fossil feet and leg bones; but they can also be seen in the back vertebrae, the length of our arms, and the position of our skulls on our backbones.
4. *Changes in social and cultural behaviour*. The main way we differ from other apes is in our social and cultural lives. This development can be followed only indirectly in fossils. Sexual dimorphism, for example, is probably related to the breeding systems. In apes other than humans, males weigh about twice as much as females on average in gorillas and orangutans, and about 1.35 times as much in chimpanzees. In humans sexual dimorphism is reduced; males on average weigh about 1.2 times as much as females. Sexual dimorphism may have reduced in our ancestors when we evolved reproductive pair bonds – prolonged pairing is found in most human societies but not in other great apes. The cultural state of a society can be followed in tools and other artifacts associated with fossils. The main innovation underlying modern human culture is language. The origin of language is hardly to study, with very indirect clues coming

from jaw and throat anatomy and from the symbolic richness of artifacts associated with fossils.

The First Humans

In 1924, British anthropologist Raymond Dart announced that a fossilized skull discovered in a South African quarry was the remains of an early human. He named his "ape-man" *Australopithecus africanus* ("southern ape of Africa") who lived between 2 and 3 million years ago in the pliocene. With the discovery of more fossils, it became clear that *Australopithecus* was in fact a hominid that walked fully erect (was bipedal) and had humanlike hands and teeth. However, the brain of *Australopithecus* was only about one-third the size of a modern human's brain. Various species of *Australopithecus* existed for over 3 million years, probably beginning about four and a half million years ago.

In 1974, in the Africa region of Ethiopia, paleoanthropologists discovered an *Australopithecus* skeleton that was 40% complete. "Lucy' as the fossil was named, was petite-only about 1 m tall, with a head about the size of a softball. The skeleton is 3.2 million years old. Lucy and similar fossils have been considered sufficiently different from *Australopithecus africanus* to be named a separate species, *Australopithecus afarensis* (for the Africa region). The fragments of the pelvis and skull bones indicate that *A. afarensis* walked on two legs (Figure 6.25). Fossils discovered in the early 1990s extend the longevity of *A. afarensis* as a species to a span of at least one million years.

Since 1994, paleoanthropologists working at other sites in East Africa have unearthed hominid fossils that predate *A. afarensis*. One called *Australopithecus anamensis*, dates from about 4 million years ago; another very different hominid, named *Ardipithecus ramidus*, probably lived about 4.4 million years ago. With the discovery of these oldest known hominids, the fossil record of humanity is creeping closer to the ape-human split estimated to have occurred 5 to 8 million years ago.

The phylogenetic relationships among early hominids are stirring much debate. One issue is whether *Ardipithecus ramidus* is the ancestor of Australopithecus or an evolutionary side branch that faded into extinction. Also at issue is the origin of bipedalism. The oldest hominid fossils challenge a long-standing hypothesis that bipedalism evolved when hominids began living on the African savanna (grasslands dotted with trees). *Australopithecus anamensis*, which was completely bipedal, apparently lived in lakeside forests. *Ardipithecus*, whose posture is still being debated, inhabited dense forests. Recent evidence also suggests that some populations of *Australopithecus afarensis* lived in wooded areas.

Another debate centers on the evolutionary fate of *Australopithecus afarensis*. For the million years that it is represented in the fossil record, *Australopithecus afarensis* changed little. Then, beginning about three million years ago, an adaptive radiation began that produced several new hominid species, including *Australopithecus africanus* and several heavier-boned

Australopithecus species. This period of speciation also produced *Homo habilis*, the first member of our genus, which debuts in the fossil record about 2.5 million years ago. Was *Australopithecus africanus* ancestral to any or all of these diverse hominids? Or was Lucy a member of a species that branched early from an unknown ancestor that also gave rise to Homo?

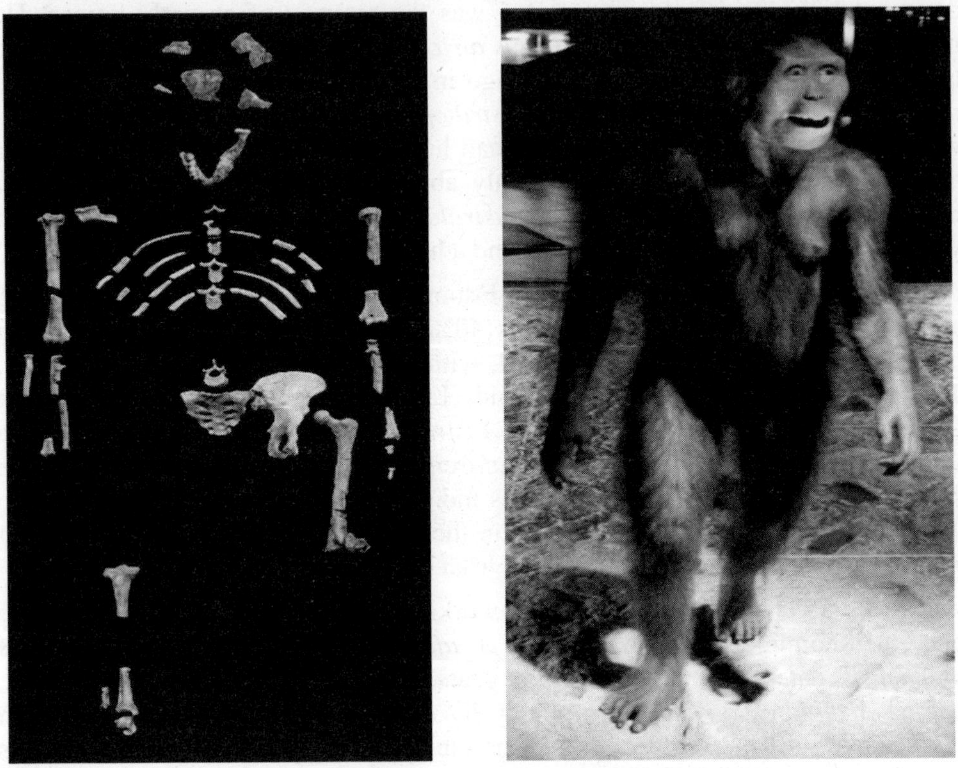

(a) (b)

Figure 6.25. ***Australopithecus afarensis***. (a) Lucy, a 3.18-million-year-old skeleton, showing fragments of pelvis and skull part of *Austraopithecus afarensis*. (b) A reconstruction of a female *Austraopithecus afarensis*.

Whatever the outcome of debates about the phylogeny of early hominids, the fossil record makes one fact of our history clear: Hominids walked upright for about two million years with no substantial enlargement of the brain. Perhaps bipedalism freed the hands for gathering food or caring for babies; the making of sophisticated tools came much later. Figure 6.26 shows one possible phylogenetic scheme that broadly traces relationship among various known hominid fossil groups beginning with *A. afarensis*.

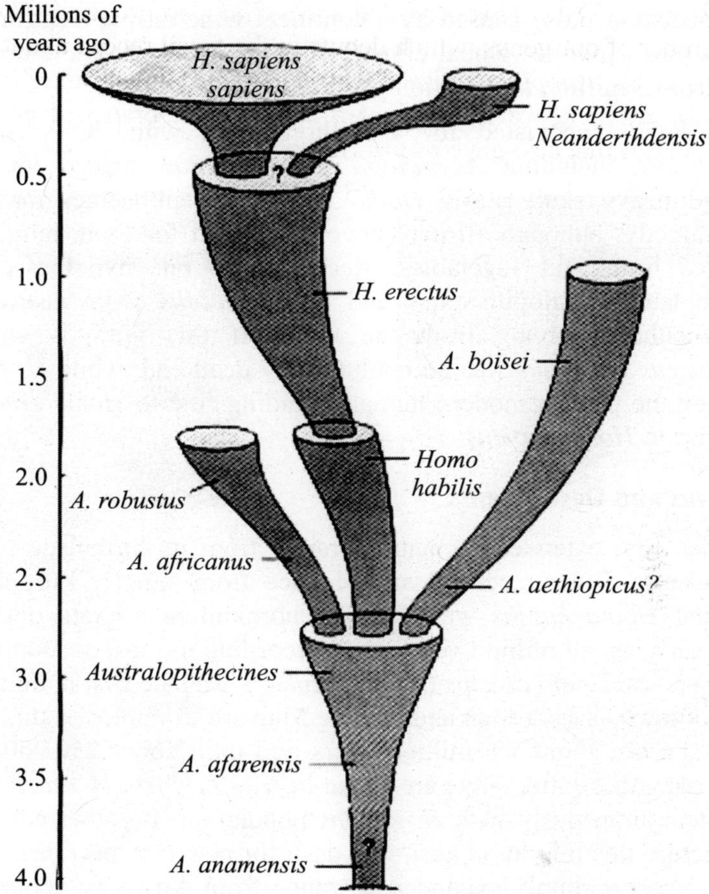

Figure 6.26. One of the possible phylogenetic scheme depicting evolutionary relationships between presently known hominid groups. This tree is based on findings of an *A. boisei* form (*A. aethiopicus*?) that indicates the *boisei* line was evolving parallel with the *A. africanus* - *A. robustus* line. Some schemes combine *A. boisei* and *A. robustus* into a single lineage and suggest only two major branches leading from *A. africanus* - one to the robust *Australopithecines* and the other to *Homo*.

6.6.4 Occurrence of hominids to be classified as genus *Homo*

Homo habilis

Enlargement of the human brain is first evident in fossils dating back to the latter part of australopithecine times, about 2.5 million years ago. Simple handmade stone tools are sometimes found with the larger-brained fossils named *Homo habilis* ("handy man"). After walking upright for at least two million years, hominids were finally beginning to use their brains and hands to fashion tools.

The hominid radiation that included *Homo habilis* is the focus of yet another important debate in paleoanthropology. One hypothesis is that the hominid radiation occurred very rapidly (from about 2.8 to 2.5 million years ago) and was part of a speciation pulse caused by a continent-wide drying trend in Africa. A related hypothesis invokes global climate change and expands the time frame to the period from 3 million to 2 million years ago.

Homo habilis coexisted for a million years with the smaller-brained *Australopithecus*, including *Australopithecus robustus*, named for its stocky features and heavy skull. Early *Homo* and australopithecines may not have competed directly, although all may have scavenged for food, hunted animals, and gathered fruits and vegetables. According to one hypothesis of human origins, the late australopithecines and *Homo habilis* were distinct lines of hominids, neither evolving from the other. If this view is correct, then *Australopithecus africanus* was an evolutionary dead end, while *H. habilis* may have been on the path to modern humans, leading first to *Homo erectus*, which later gave rise to *Homo sapiens*.

Homo erectus and Descendants

What species first extended humanity's range from its birthplace in Africa to other continents? Some recent fossil evidence from sites in Europe and Asia indicates that *Homo habilis* was the first hominid to migrate out of Africa, perhaps as early as 1.8 million years ago. According to most accounts, however, a younger species, *Homo erectus* ("upright man"), was the first hominid to do so. The fossils known as Java Man and Beijing Man are examples of this species. *H. erectus* lived from about 1.8 million years ago until about 250,000 years ago. Fossils covering that entire range are found in Africa, where *H. erectus* continued to live contemporaneously with *H. erectus* populations on other continents. We need not picture this migration as a mad dash for new territory, or even a casual stroll. If *H. erectus* simply expanded its range from Africa by about a mile per year, it would take only about 15,000 years to reach Java and other parts of Asia and Europe. The gradual spread may have been associated with a change in diet to include a larger proportion of meat. In general, animals that hunt require more geographical territory than animals that feed exclusively on vegetation.

Homo erectus was taller than *H. habilis* and had a larger brain capacity. During the 1.5 million years the species existed, the *H. erectus* brain increased to as large as 1200 cc, a brain capacity that overlaps the normal range for modern humans.

The intelligence that evolved during the African origins of *H. habilis* enabled these humans to survive in the colder climates of the north once migration began. *Homo erectus* resided in huts or caves, built fires, clothed themselves in animal skins, and designed stone tools that were more refined than the tools of *Homo habilis*. In anatomical and physiological adaptations, *H. erectus* was poorly equipped for life outside the tropics but made up for the deficiencies with intelligence and social cooperation.

Some African, Asian, European, and Australasian (Indonesia, New Guinea, and Australia) populations of *H. erectus* gave rise to regionally diverse descendants that had even larger brains. Among these descendants of *H. erectus* were the Neanderthals, who lived in Europe, the Middle East, and parts of Asia from about 130,000 years ago to about 35,000 years ago. (They are named Neanderthals because their fossils were first found in the Neander Valley of Germany.) Compared with us, Neanderthals had slightly heavier brow ridges and less pronounced chins, but their brains, on average, were slightly larger than ours. Neanderthals were skilled toolmakers, and they participated in burials and other rituals that required abstract thought. Much current research on Neanderthal skulls addresses an intriguing question: Did Neanderthals have the anatomical equipment necessary for speech?

The oldest known post — *H. erectus* fossils, dating back over 300,000 years, are found in Africa. Many paleoanthropologists group these African fossils along with Neanderthals and various Asian and Australasian fossils as the earliest forms of our species, *Homo sapiens*. These regionally diverse descendants of *H. erectus* are sometimes referred to as "archaic *Homo sapiens*."

6.6.5 The Emergence of anatomically modern humans: *Homo sapiens*

Paleoanthropologists refer to the kind of human beings that live today – *Homo sapiens* – as "anatomically modern humans." Anatomically modern humans differ from the fossils discussed so far in a series of details in skull anatomy. Our brains are a different shape from *H. erectus*, and our faces are flatter. When and where did anatomically modern humans originate? For the past 15 years, this question has been the topic of a debate between two hypotheses (theories) (Figure 6.27). By 500,000 years ago, human populations descended from *H. erectus* were established in Asia (and Australia) and Europe as well as Africa. The taxonomic names of these regional forms are not agreed on. Some taxonomists refer to all the regional forms as "archaic *Homo sapiens*"; some classify different regional forms as different species (*H. neanderthalensis* in Europe, for example); and some call them subspecies of *H. erectus*. In the view of some paleoanthro-pologists, these archaic *Homo sapiens* gave rise to fully modern humans. According to this model, called the multiregional model, modern humans evolved in parallel in different parts of the world. If this view is correct, then the geographic diversity of humans originated relatively early, when *Homo erectus* spread from Africa into the other continents between 1 and 2 million years ago. This model accounts for the great genetic similarity of all modern people by pointing out that occasional interbreeding among neighboring populations has always provided corridors for gene flow throughout the entire geographical range of humanity.

A number of paleoanthropologists, including Christopher Stringer of the University College, London, interpret the fossil record differently and have proposed an alternative to the multiregional model. The debate has focused partly on the relationship between Neanderthals and the modern humans of Europe and the Middle East. Perhaps the most famous fossils of modern *Homo sapiens* — skulls and other bones that look essentially like those of today's humans—are the remains of Cro-Magnons. Named for the French cave in which these fossils were first discovered, Cro-Magnons date back about 35,000 years. But the oldest fully modern fossils of *Homo sapiens*, about 100,000 years old, are found in Africa; similar fossils almost as ancient have also been discovered in caves in Israel. The Israeli fossils were found not far from other caves containing Neanderthal-like fossils ranging in vintage from about 120,000 to 60,000 years old. These two human types apparently coexisted in this region for at least 40,000 years, from the time modern forms appeared 100,000 years ago. And, according to Stringer and his supporters, the persistence of distinct Neanderthal and modern forms during their time of coexistence means that these two human types did not interbreed. If this interpretation of the Israeli fossils is correct, then the Neanderthals of that region could not have been the ancestors of the modern humans that also lived there. In fact, based on the Israeli bones and other fossil data, Stringer postulates that Neanderthals and the other so-called archaic *Homo sapiens* outside of Africa were evolutionary dead ends.

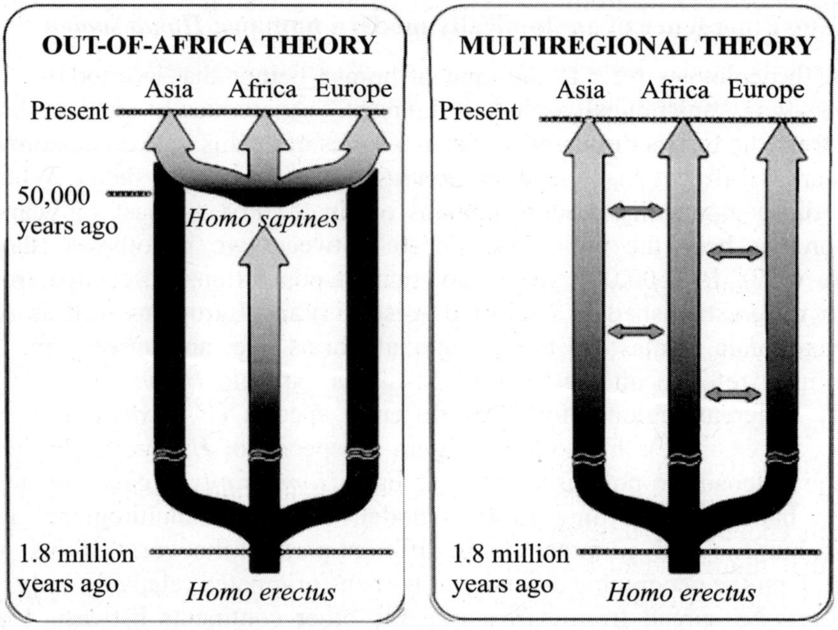

Figure 6.27. Representation of the two theories for the origin of anatomically modern humans

Thus, Stringer is among the anthropologists who contend that modern humans evolved from Homo erectus first in Africa, then migrated to other continents, replacing Neanderthals and the other regional descendants of *H. erectus*. This model, called the "Out of Africa" or Monogenesis model, contrasts sharply with the Multiregional model. According to the Monogenesis model, modern humanity did not emerge in many different parts of the world, but only in Africa, from which modern humans dispersed relatively recently. If this view is correct, then the geographical diversification of modern humans occurred within just the past 100,000 years (compared to at least 1 million years for the Multiregional model). To Stringer, the strongest argument for an exclusively African genesis for modern humans is that only African fossils chronicle the complete transition from archaic *Homo sapiens* to modern humans.

In the late 1980s, Rebecca Cann and other geneticists in the laboratory of the late Allan Wilson at the University of California, Berkeley, made news with research results that seemed to support the Monogenesis model of human origins. Instead of digging in the field for bones, Wilson's group probed in the DNA of living humans for clues about our ancestry. Specifically, these scientists compared the mitochondrial DNA (mtDNA) of a multiethnic sample of more than 100 people representing four continents. The greater the difference between the mtDNA of two people, the longer ago that mtDNA diverged from a common source. Using computers to analyze their data, the geneticists traced the source of all human mtDNA back to Africa. The surprise was the calculation that the divergence of mtDNA from this common source began just 200,000 years ago, much too late to represent the dispersal of *Homo erectus*. Advocates of the monogenesis model cheered the genetic evidence for a second, much later dispersal of Africans in the form of modern humans.

However, in spite of its advantages, mtDNA acts as only a single genetic unit whose genealogy may not necessarily coincide with other genetic units. Other studies have therefore broadened the data using information on many nuclear genes. Nei and Roy Chodhury, for example, calculated genetic distances between 26 human populations for 29 different nuclear genes (see Figure 6.28), and also suggest a single African origin for *Homo sapiens* with subsequent widespread geographic divergence (Figure 6.29).

In 1992, several researchers challenged the Berkeley group's interpretation of the mtDNA data. Questions were raised on the methods used to construct phylogenetic trees from the mtDNA comparisons and the reliability of changes in mtDNA as a molecular clock for timing branch points in these trees. The criticism encouraged proponents of the multiregional model, who continue to argue that multiregional evolution of modern humans fits the fossil evidence better than does African monogenesis. These scientists interpret certain fossils in various parts of the world as links between the regional versions of archaic *Homo sapiens* and the current people native to those continents.

The debate about the origin of modern humans continues, largely fueled by molecular studies. Strengthening the case for a spread of modern humans out of Africa, one research group has found greater genetic diversity within African populations south of the Sahara than in other parts of the world. The result is predicted by the hypothesis that modern humans originated in southern Africa and therefore have the longest history of genetic diversification in that region. If populations in other regions began by later migration, the founder effect and genetic drift could account for the smaller genetic diversity.

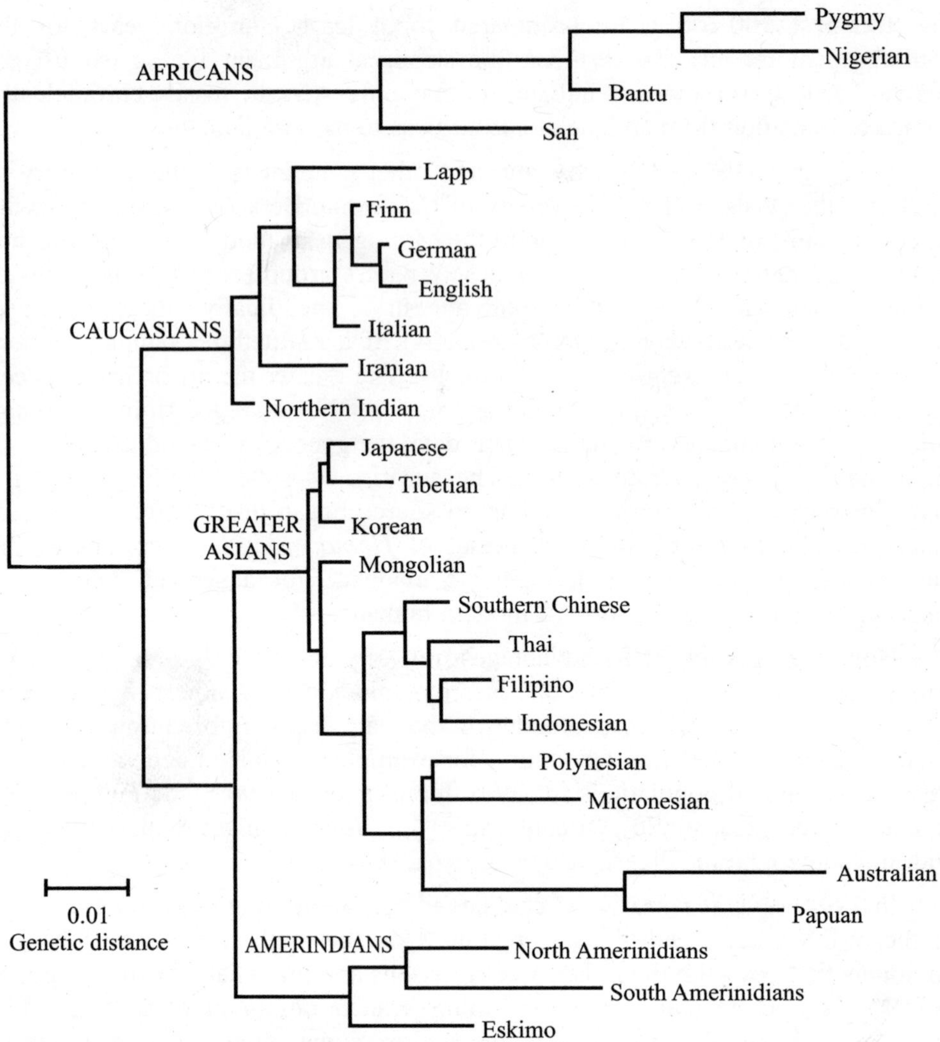

Figure 6.28. A phylogenetic relationship among 26 human populations based on the genetic distance measurement on 29 genetic loci.

In a more recent study, molecular systematists obtained a sample of mtDNA from an unusually well-preserved Neanderthal fossil in Europe. After a painstaking effort to eliminate contamination of the mtDNA with DNA from other sources, they used polymerase chain reaction (PCR) to amplify the DNA traces and produced enough for comparison to the DNA of living humans. Their analysis indicated that the Neanderthal individual was genetically very different from modern Europeans. This result supports the monogenesis model, but generalizations cannot be made from one specimen. However, studies of DNA from the Y chromosome of modern-day humans also provide support for the out of Africa theory. In contrast, other research points to an archaic human population in Asia as the source of genes coding for some blood proteins.

Perhaps, as some researchers have proposed, an intermediate model for the origin of modern humans is more realistic than the monogenesis and multi-regional models: Modern *H. sapiens* may be the result of a migration out of Africa as well as some genetic contribution from non-African archaic groups.

6.7 CULTURAL EVOLUTION

An erect stance was the most radical anatomical change in our evolution; it required major remodeling of the foot, pelvis, and vertebral column. Enlargement of the brain was a secondary alteration made possible by prolonging the growth period of the skull and its contents. The brains of nonprimate mammalian fetuses grow rapidly, but the growth usually slows down and stops not long after birth. The primate brain continues to grow after birth, and the period of growth is longer for a human than for any other primate. The extended period of human development also lengthens the time parent's care for their offspring, which contributes to the child's ability to benefit from the experiences of earlier generations. This is the basis of culture — the transmission of accumulated knowledge over the generations. The major means of this transmission is language, written and spoken.

Cultural evolution is continuous, but there have been three major stages. The first stage began with nomads who hunted and gathered food on the African grasslands 2 million years ago. They made tools, organized communal activities, and divided labor. The second stage came with the development of agriculture in Africa, Eurasia, and the Americas about 10,000 to 15,000 years ago. Along with agriculture came permanent settlements and the first cities. The third major stage in our cultural evolution was the Industrial Revolution, which began in the eighteenth century.

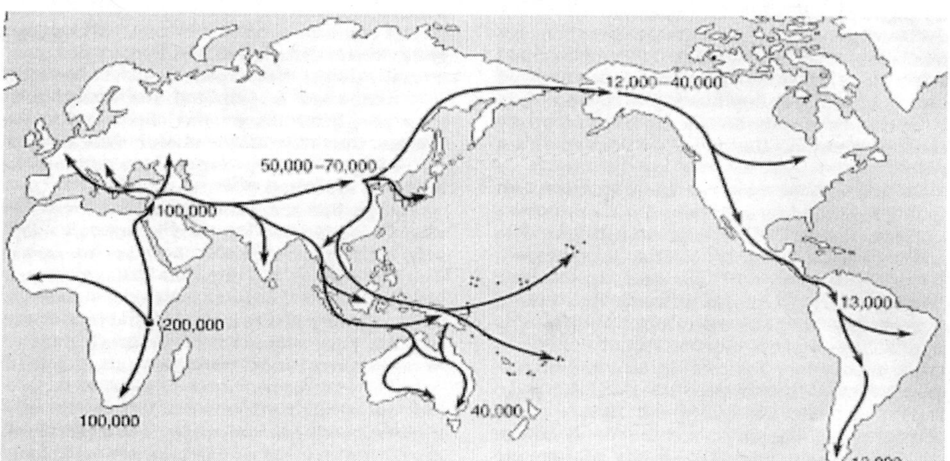

Figure 6.29. A proposed scenario for the geographical distribution of *Homo sapiens* from their African origin. The number derives from a study of genetic distances between 26 human populations and represent estimate dates at which these population reached their destinations

Evolution of the human brain may have been anatomically simpler than acquiring all stance, but the consequences of cerebral expansion have been enormous. Cultural evolution made *Homo sapiens* a new force in the history of life—a species that could defy its physical limitations and shortcut biological evolution. We do not have to wait to adapt to all through natural selection; we simply change the environment to meet our needs. We are the most numerous and widespread of all large animals, and everywhere we go, we bring environmental change. We are changing the world faster than many species can adapt; the rate of extinctions in the 20th century is 50 times greater than the average for the past 100,000 years.

This rapid rate of extinction is mainly a result of habitat destruction, which is a function of human cultural changes and overpopulation. Feeding, clothing, and housing of billions of people who now exist imposes an enormous strain on earth's capacity to sustain life. If all these people suddenly assumed the high standard of living enjoyed by many people in developed nations, it is likely that earth's support systems would he overwhelmed. Already, for example, current rates of fossil-fuel consumption, mainly by developed nations, are so great that waste carbon dioxide may be causing the temperature of the atmosphere to increase enough to alter world climates. Today, not just individual species, but entire ecosystems, the global atmosphere, and the oceans, are seriously threatened. Tropical rain forests, which play a vital role in moderating global weather, are being cut down at a startling rate. Scientists have hardly begun to study these ecosystems, and many species in them may become extinct before they are even discovered.

Of the many crises in the history of life, the impact of one species, *Homo sapiens*, is the latest and potentially the most devastating.

Chapter 7

Origin and Evolution of Angiosperms

7.1 Introduction: An Overview of Evolution of Land Plants

7.2 Fossil evidences of Angiosperms

 7.2.1 The Koonwara Angiosperm

 7.2.2 Features of other fossil flowers

7.3 The Origin and Radiation of Angiosperms

 7.3.1 Time period of Angiosperms evolution

 7.3.2 Probable site of Angiosperms evolution

 7.3.3 Causes of Angiosperms evolution and diversification

7.4 Evolutionary Transformation of a Gymnosperm into a Flowering Plant

7.5 Angiosperm Phylogeny

 7.5.1 The basal angiosperms

7.1 INTRODUCTION: AN OVERVIEW OF EVOLUTION OF LAND PLANTS

The terrestrial surface was first colonized by microbes. Fossil record of 1000 million years ago suggests the existence of prokaryotic cells in terrestrial environment. Terrestrial life existed only in microbial form until the land was colonized by plants and fungi, but for plants we have widely accepted evidence of fossil spores of 475 million years ago. Land plants are most closely related to a group of green algae called Charophyceae. The major events in the early evolution of land plants were the evolution of a resistant spore stage, then the evolution of vascular tissue, followed by roots and then leaves. Resistant spore stage is evident in Bryophytes but they lack vascular tissues. Pteridophytes such as ferns have vascular tissues. Vascular tissues having lignin containing tracheid cells helps plant to support itself on land.

About 430 million years ago the earliest fossil of whole plant known as *Clarksonia* was found. These early land plants lacked roots and leaves and simply had branching stems which performs photosynthesis as stomata are visible in fossil stems. Fossils with leaves appear 390-350 million years ago. The evolution of leaves coincides with fall of atmospheric carbon dioxide concentration upto 90% at that time. This reduction in atmospheric carbon dioxide level was not only by relatively minor photosynthetic activity of stem of early evolved land plants but more importantly through the evolution of roots. Roots promote weathering which in turn removes large amounts of carbon dioxide from the atmosphere. This fall in carbon dioxide level set up a selection pressure in favour of leaves which results in evolution of leaves and more powerful photosynthesis, further reduced the carbon dioxide levels.

After vascularization in land plants, seed plants emerged as main land groups. Fossil seeds exist in the Carboniferous period when the coal deposits were formed. However, seed plants were a minor group at that time. Coal is mainly formed from fossil pteridophytes. The two groups of seed plants – Gymnosperms and Angiosperms proliferated later. Fossil evidences suggest that the Gymnosperms originates around 360 million years ago in late Devonian period. The earliest known fossil seeds (about 360 million years ago) were described as a loose cluster of seeds, each consisting of a megasporangium surrounded by a protective, integument like layer with several finger like projections (Fig. 7.1).

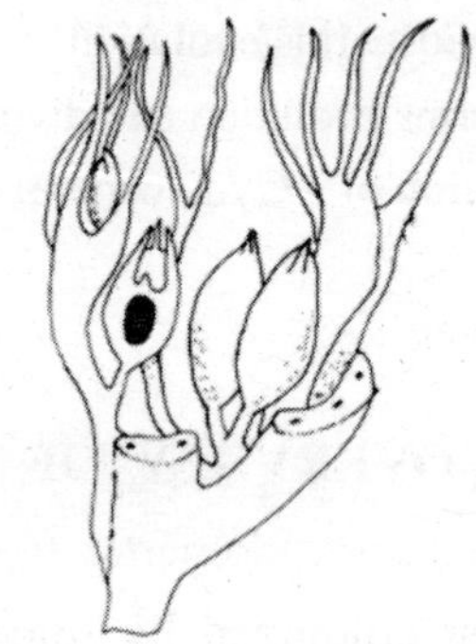

Fig. 7.1. Reconstruction of ovules (with finger like projections) of a plant that grows in the late Devonian period (about 360 million years ago)

The angiosperms, or flowering plants, are one of the major groups of extant seed plants and arguably the most diverse major extant plant group on the planet, with at least 260,000 living species classified in 453 families. Flowering plants are more recently evolved and most derived of all plant groups. Flowering plants are the most successful of all plant groups in terms of their diversity and are classified together in a single division *Magnoliophyta*, also known as division *Anthophyta*. They represent one of the greatest terrestrial radiations and are product of a long line of evolutionary development that culminated in the highly specialized organs for reproduction that we know as the flower. Angiosperms are the dominant plant life of the geological era in which we live.

The first whispers of angiosperm trait appear in the fossil record of the early Cretaceous period, around 130-136 million years ago, followed by a rise to ecological dominance in many habitats before the end of the Cretaceous. However, RNA and DNA sequence suggest the lineage giving rise to angiosperms may have split from the other seed plants as early as 280 million years ago. This molecular evidence is supported circumstantially by the chemical analyses of the rock strata, revealing the presence of "oleanane" in rock deposits dating from 290 to 235 million years ago. Oleanane is a compound produced only by modern angiosperms that acts as a defense against insects.

They occupy every habitat on earth except extreme environments such as the highest mountaintops, the regions immediately surrounding the poles, and the deepest oceans. They live as epiphytes (i.e., living on other plants), as floating and rooted aquatics in both freshwater and marine habitats, and as terrestrial plants that vary tremendously in size, longevity, and overall form. They vary in their size from smallest plants (Wolffia) to some of the tallest (Eucalyptus) (Fig. 7.2).

Figure 7.2. (a) *Wolffia*, the smallest angiosperm, (b) *Eucalyptus*, one of the tallest flowering plant

There is a huge amount of diversity in chemistry (often as a defense against herbivores), reproductive morphology, and genome size and organization that is unparalleled in other members of the Plant Kingdom. Furthermore, angiosperms are crucial for human existence; the vast majority of the world's crops are angiosperms, as are most natural clothing fibers. Angiosperms are also sources for other important resources such as medicine and timber.

Charles Darwin once remarked that the sudden and abundant appearance of angiosperms in the fossil record as "an abominable mystery" because very little was known about the origin and rapid diversification of angiosperms into diverse group. Botanists have used several different kinds of evidence to try to solve some of the mysteries associated with evolution of flowering plants, including evidence from the fossil record, comparative morphology and from gene sequencing. The contribution of such evidence to studies of the origin and evolution of angiosperms is the central theme of this chapter.

7.2 FOSSIL EVIDENCES OF ANGIOSPERMS

Fossils are considered to be the best evidence for the life forms which existed in past, they provide a basis for some general ideas about the origin and evolution of life forms. In case of plants the fossil record of plants may be 1% complete and that as much as 90% of the plant species that ever existed in the past decomposed without leaving a trace of their existence. Therefore, little is known about the plant species which ever existed in the past and are now extinct. In the context to angiosperms most of the oldest fossils are so fragmented and incomplete that paleobotanists are not certain that they are angiosperms at all. Still one particular fossil consisting of all the parts of the flower attached to a reasonably intact plant was found near Koonwarra, Australia. This fossil of the flowering plant was 120 million years old and paleobotanists believe that this plant represents the ancestral type of flower.

The earliest leaf fossils definitely considered to be those of angiosperms are from the lower early Cretaceous period, about 130 million years ago. These leaf fossils represents both dicot and monocot leaves, for example *Magnoliaephyllum*, *Ficophyllum* (Fig. 7.3). The oldest wood that seems to be from an angiosperm comes from the Aptian Epoch (125 million years ago) of Japan.

7.2.1 The Koonwarra Angiosperm

The Koonwarra angiosperm, the earliest flower fossil was discovered in 1986, and it was thought to be a fern, but closer examination revealed that it was Carpel bearing inflorescences (Fig. 7.4). The entire plant was near about 3 cm long and it closely resembles a black pepper plant. The Koonwarra angiosperm resembles in many features to many present-day angiosperms, for example it had small flowers without petals, a spike-like inflorescence, single carpel ovaries with short stigmas lacking styles, and imperfect flowers with several bracts at their bases. These features are common in many modern angiosperms families like Saururaceae, Piperaceae, and Chloranthaceae, all of which are dicots (Fig. 7.5). further the venation pattern of leaf of Koonwarra angiosperm resembles present day families of Aristolochiaceae, Smilacaceaea and Dioscoreaceae, the latter two being families of monocots.

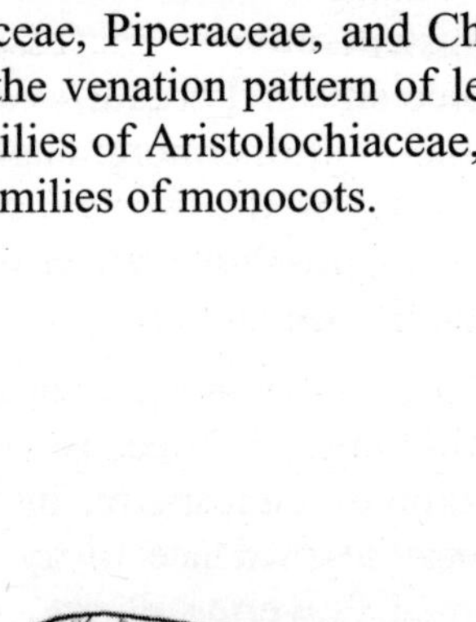

Figure 7.3. A leaf of Magnoliaephyllum showing similarities to the leaves of living Magnolias

Figure 7.4. Reconstruction of oldest known fossil Hower, which is about 120 million years old, was found in the Koonwara fossil beds near Melbourne, Australia

Figure 7.5. *Saururus* sp., Saururaceae (commonly known as Lizard's tail)

The Koonwarra angiosperm gives an idea about how the ancestors of flowering plants may have looked: a small, rhizome-bearing herb that had secondary growth, small reproductive organs and simple, imperfect flowers with complexes of bracts at their bases. All the families of living flowering plants which share these features with Koonwarra angiosperm are believed to be primitive members of the dicots and monocots. Furthermore, the appearance of this plant near the apparent beginning of the evolution of angiosperms and its similarity to dicots and monocots suggest that the Koonwarra angiosperm evolved before the divergence between monocots and dicots. This means that the evolutionary lineage of monocots and dicots gets separated in less than 120 million years ago, probably from the ancestor similar to the Koonwarra angiosperm.

7.2.2 Features of other fossil flowers

Flowering plants were evolving rapidly during the time that the Koonwarra angiosperms lived. This rapid diversification of angiosperms is evident and based on fossil evidences collected from different places of the world that show a variety of floral and vegetative features among different plants. These fossils represent how flowering plants may have evolved from their earlier ancestors into the vast array of species that live today. Some of such evolutionary development are described below.

Diversity in the floral features have been seen in the flower fragments of angiosperms from the early Cretaceous period. Some fossil floral parts are arranged spirally on the floral axis, as in the flower and fruits of modern *Magnolia* (Fig 7.6).

Figure 7.6. The earliest flower had floral parts arranged in spiral (like carpels of *Magnolia*)

Figure 7.7. Floral parts arranged in whorls like most extant angiosperms (Lily)

Other fossils have some or all of their floral parts arranged in circles or whorls, around the floral axis, which is the most common type of pattern found in extant angiosperms (Fig. 7.7).

Some early fossils show presence of sepal and petals, while others lack a distinct calyx and corolla. All the early flowers were radially symmetrical, and their petals were free i.e. polypetalous condition (unattached to each other Fig. 7.8). The earliest gynoecia had free carpels, and the first fruits derived from them were apparently follicles or nutlets. Fossils of gynoecia with fused carpels are not as old as those with free carpels. The first gynoecia with fused carpels may have developed into capsules. Fused carpels in more recent fossils show the development of nuts, drupes, berries and pods.

Figure 7.8. Polypetalous condition in earliest flower like *Ranunculus*

7.3 THE ORIGIN AND RADIATION OF ANGIOSPERMS

The Koonwarra angiosperm is the oldest known fossil flower but it is not the oldest known angiosperm fossil. Fossil pollen from the early Cretaceous period, perhaps 10 million years older than Koonwarra angiosperm, may have come from angiosperms. Nobody knows just how long ago the first angiosperm lived, thus we lack direct evidence for the earliest angiosperms. Many suggestions have been made from the available fossil evidences regarding the time period, place, how angiosperms evolved and what were the first angiosperm.

7.3.1 Time period of Angiosperms evolution

The sudden appearance of diversity of angiosperms in the Cretaceous period suggests that the evolution of flowering plants began much earlier, perhaps as much as 100 million years before the oldest known angiosperm fossil. If it is true, then the beginning of angiosperms may be found among plants that are believed to be more like the ancestors of angiosperms. For finding the possible ancestory and lineage to the angiosperms, paleobotanists have often focused on comparing the reproductive features of angiosperms with those of different kinds of non-flowering plants.

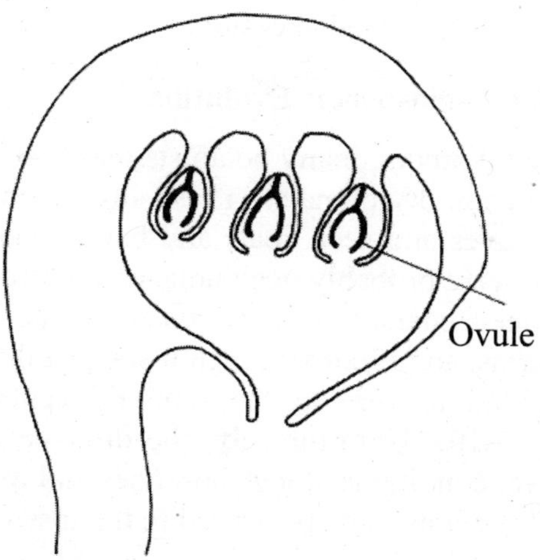

Figure 7.9. Diagram of a sectioned *Caytonia* cupule containing three seeds

To explain when angiosperms evolved, focus was given in determining when the carpel arose. One hypothesis suggests that the carpel developed from the cupule of a seed fern (pteridosperms) like *Caytonia* (Fig. 7.9). According to this hypothesis, several features of seed ferns like reticulate venations, monopodial branching, presence of cambium, presence of microsporophylls and megasporophylls on the same plant, development of triploid endosperm as extreme reduction of female gametophyte, similarities of seed structure and existence of one or several ovules surrounded by cupule tissue brings seed ferns closest to angiosperms. This hypothesis suggests that cupule tissue surrounding the ovules fused to form a closed carpel. Seed ferns were prominent in the Carboniferous period, but few persisted into the Mesozoic era. This means that the carpel, or a precarpel, may have originated as early as 200 million years ago. Another hypothesis is that the carpel arose from the longitudinal folding of an ovule-bearing leaf, i.e. from a folded megasporophyll.

Cycadeoids were also considered to be the ancestors of angiosperms, partly because the microsporangia and ovules of cycadeoids occur in the same cone. Such an arrangement is found in perfect flowers, i.e. flowers with both stamens and carpels on the same receptacle, which are the most common type of flowers in angiosperms. Further the ovules of cycadeoides are believed to have contained linear tetrads in the megasporangia, which is characteristic of angiosperms.

Recently, botanists have used the methods of cladistics to show how angiosperms may have descended rather directly from seed-fern ancestors in a line parallel to cycadeoids. At present, many paleobotanists and taxonomists believe that this transition from gymnosperm to angiosperm occurred during the Jurassic and Lower Cretaceous periods of Mesozoic era.

The evolutionary relationship between angiosperms and gymnosperms has yet to be precisely determined. Molecular data supports the hypothesis that the Bennettitales (an extinct group of cycad-like plants), the Gnetales, and angiosperms all had closely related ancestors.

7.3.2 Probable site of Angiosperms Evolution

There is a general belief among many botanists that Pre-Cretaceous angiosperms were well adapted to cool, dry climates. These angiosperms were probably small in size, with tough leaves and seed coats and having vessels in their secondary xylem. Most of them were probably deciduous and thus avoided seasonal drying. These above mentioned characters were based on the fossils of more recent, Cretaceous angiosperms, and if correct, then it suggest that the most likely places for angiosperms evolution were in the semiarid central regions of Western Gondwanaland (Fig. 7.10). Unfortunately, the drier conditions in these upland regions, unlike the wet conditions along shorelines and in lowland basins, did not allow much chance for plants to be preserved in the fossil record.

Gymnosperms were the dominant flora of lowland basins from the Jurassic and Triassic periods, but angiosperms apparently began to invade these areas by the early Tertiary period, less than 65 million years ago. The climatic and geological changes at the end of the Mesozoic era lead to invasion of angiosperms into these lowland areas. At that time, the average temperature over the earth dropped as much as 20°C. Also, mountains were forming rapidly, and continents were breaking apart and drifting northward. These drastic changes were responsible for extinction of many organisms and at the same time provided new opportunities for the migration and evolutionary radiation of angiosperms into lowland habitats. Thus, although the first angiosperms probably evolved in cool, dry climates of uplands sometimes before the Cretaceous period, the explosive diversification of flowering plants seems to have occurred much later in Tertiary lowlands.

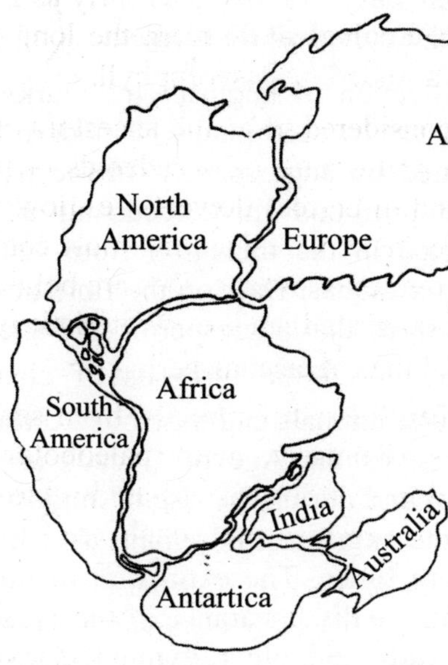

Figure 7.10. Positions of the earth's land masses in the early Triassic period

7.3.3 Causes of Angiosperms' evolution and diversification

Angiosperm evolution and diversification has often been linked to the diversification of pollen and nectar-collecting insects on the supposition that insect pollination provides new possibilities for reproductive isolation and therefore elevation of speciation rates. Compared to wind pollination, which is widespread in other seed plants, insect pollination may also permit effective outcrossing at lower population densities and in a greater range of environments, perhaps thereby reducing extinction rates. These factors may explain the diversity of some of the most speciose angiosperm families such as orchids.

The earliest insects that could have been pollinators were probably beetles. Cycadeoids were already specialized for pollination by beetles long before the appearance of angiosperms in the fossil record. Cycadeoids had tiny ovules that were protected by scales, which is a feature that is consistent with pollination by insects with chewing mouthparts, such as beetles. By analogy with cycadeoids, early angiosperms may therefore adapted pollination by beetles which had previously established relationships with cycadeoids and other seed plants. The coevolutionary interaction between flowering plants and pollinators begins and in between the late Cretaceous and early Tertiary period with the rise of specialized pollinator groups like lepidopterans (e.g., butterflies, moths) and hymenopterans (e.g., bees) further leads to rapid diversification of angiosperms. The rise to dominance by angiosperms in the Tertiary period, therefore, seems to have been greatly influenced by adaptations for pollination by an increasing diversity of flying insects. Thus, insects probably played an important role in the evolution of angiosperms into the largest and most diverse group of land plants.

7.3.4 Rise of Angiosperms to Ecological dominance

The Mid-Cretaceous diversification of angiosperms marks the transition from Mesozoic ecosystems dominated by seed plants to modern Late Cretaceous and Tertiary ecosystems dominated by angiosperms. In the palaeobotanical record this change is much more profound than occurring at Cretaceous-Tertiary boundary and is also reflected in the transition from sauropod dominated to ornithopod dominated dinosaur faunas. Data on the floristic composition of both macrofloras and microfloras show that angiosperms had attained diversity levels of about 50-80% by the end of the Cretaceous period.

Dinosaurs were spectacular animals and they lived throughout the Mesozoic era in the Age of Dinosaurs, or the Age of Cycads (the term preferred by botanists for the same period) and during the rise of angiosperms. Of these three groups of organisms, only the angiosperms remain as a highly diversified and dominant life form in modern times. The extinction of dinosaurs by the early Tertiary period coincides with the disappearance of the cycadeoids. By this time, the number of different cycads and other gymnosperms had also begun to diminish. Recent studies by paleobotanists and paleontologists have focused on how the gymnosperms, angiosperms, and dinosaurs may have interacted and even influenced their respective evolutionary histories and ecology of the area.

The main question asked by botanists is, "How did angiosperms rise to ecological dominance at the expense of gymnosperms?" One possible explanation involves adaptations to insect pollination by angiosperms, which has already been discussed and another explanation involves dinosaurs.

Diplodocus, *Brachiosaurus*, and *Apatosaurus* (popularly known as Brontosaurus) were the largest herbivores ever to walk the earth. These animals probably roamed in great herds, browsing on the tops of coniferous trees and other large gymnosperms. This type of browsing would not have killed large trees. Further these large herbivores did not destroy seedlings by browsing, because probably they could not reach down to them. Thus, gymnosperms could flourish without being destroyed by large herbivores.

At the beginning of the Cretaceous period the coexistence of the large herbivores dinosaurs and tall gymnosperms began to disintegrate probably due to beginning of dinosaurs mysterious shrinkage in body size. By the late Cretaceous period, large herbivores had been replaced by smaller, lower browsers of the ornithischian type (e.g., *Parasaurolophus*, *Montanoceratops* and *Ankylosaurus*). These dinosaurs had relatively large, muscular heads with flat, grinding teeth for chewing plant tissues. Although the Jurassic period, when the largest dinosaurs lived, is known as the Age of Dinosaurs, there were actually more kinds of dinosaurs towards the end of the Cretaceous period. The greater diversity of smaller, late Cretaceous dinosaurs may indicate a higher degree of specialization in plant – dinosaur relationships in earlier period.

The decrease in body size over a long period of time and increase in diversity of the dinosaurs coincided with the evolutionary radiation of angiosperms and the decline of gymnosperms. Gymnosperms were probably destroyed by the lower browsers, which would have eaten small seedlings before they were able to reach maturity and produce seeds. Since the first angiosperms were smaller and herbaceous, they probably grew and reproduced more rapidly than woody gymnosperms, and therefore had better chance of producing seeds before being eaten. Furthermore new habitats were opened for angiosperms due to the devastation of gymnosperms by small dinosaurs, these new habitats were invaded by the angiosperms and results into evolutionary radiations of angiosperms.

While the dinosaur-gymnosperm-angiosperm connection seems plausible, the suggestion that angiosperms diversified by adapting to insect pollination is equally so. It may be that each of these explanations is correct and that dinosaurs and insects both influenced the rapid evolution of angiosperms in the Cretaceous period. However, the greatest diversification of angiosperms occurred in the tertiary period, after the dinosaurs were already extinct.

7.4 EVOLUTIONARY TRANSFORMATION OF A GYMNOSPERM INTO A FLOWERING PLANT

The evolutionary transformation of a gymnosperm into a flowering plant was not a simple process and it involved many alterations. The most important was the conversion of gymnosperm sporophylls into stamens and carpels, resulting in the formation of flowers. In gymnosperms, microspores and megaspores are formed an open sporophylls in male and female cones respectively. In angiosperms, however microsporangia sporophylls evolved into closed anther containing pollen whereas megasporangia sporophylls evolved into closed carpel (angiospermous) containing eggs. Because gymnosperms have rather flat, leaflike sporophylls, those flowering plants that have similar type of reproductive structures are considered to be the most relictual. For example, groups such as *Austrobaileya*, *Magnolia*, *Himantandra*, and *Degenaria* have flat stamens without distinct filament and anther portions, these features have not undergone much evolutionary change, so they are still similar to the ancestral features. The stamens (microsporophylls) of these four living dicots are much fleshier, flatter, and more leaflike than most other stamens of other living species of angiosperms. Also, the sporogenous tissue is rather massive with prominent internal masses, similar to that of cycads and cycadeoides (Fig. 7.11).

Carpels in *Austrobaileya*, *Magnolia*, and *Degenaria* are similarly leaflike, resembling young leaves whose blades have not yet opened. In these carpels, the ovary edges have rows of hairs that function as a stigmatic surface instead of having stigma and style (Fig. 7.12). Whereas ovules (megasporangia) in gymnosperms are exposed on the megasporophyll surface, the folded nature of the sporophyll in flowering plants offers the ovule some protection.

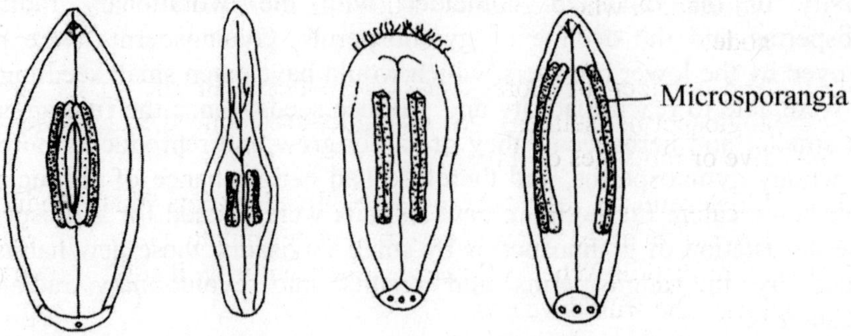

Figure 7.11. The stamens (Microsporophylls) of four living dicots

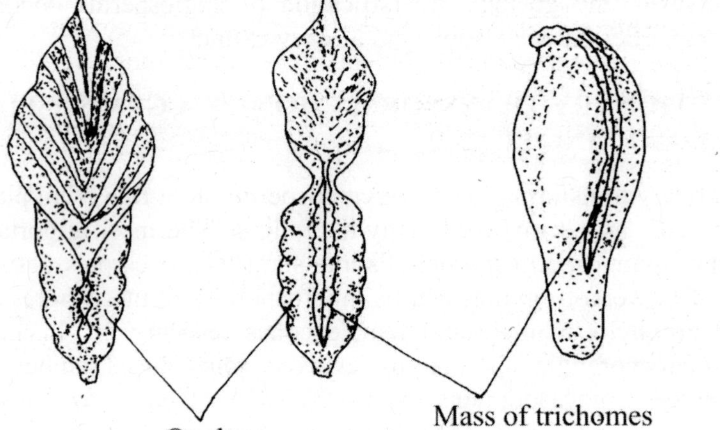

Figure 7.12. The carpels (megasporophylls) of living Dicot

The transition from gymnosperms to angiosperms seems likely as a result of major changes and adaptations that occurred within gymnosperm lineage:

1. The process of double fertilization was probably the first adaptation because it is universally present in all flowering plants and unique to the angiosperms. Double fertilization is a process in which each pollen grain produces two sperms; one fuses with an egg to form the zygote, and the other fuses with one or more polar nuclei in the female gametophyte to form an endosperm (nutritive material that surrounds the embryo within the seed).
2. Stamens and carpels become less leaf-like, with carpels often fusing into compound carpels.

3. Floral parts evolve from being spiral-like cone scales to being whorled. A whorl consists of three or more parts attached to the same node.
4. The number of floral parts is reduced. In the major groups of angiosperms, stamens, for example, are reduced from many to four, five or multiples of three.
5. Indeterminate leafy shoots (long-lived shoots that produce only leaves) evolved into short-lived determinate shoots modified for reproduction. When a flower reaches maturity, it stops growing, seeds form, and fruit develops.
6. Within the vegetative body, the major transitions were the evolution of vessel elements and sieve tubes. In living plants, vessel elements that most closely resemble tracheids are long and narrow while wide vessels elements are derived. Most long, narrow vessel elements have scalariform perforation plates, so it clearly indicates that these vessel elements evolved from tracheids with scalariform pits. Similarly, a sieve element conducts photosynthate mostly longitudinally, it is most efficient if the sieve areas and sieve pores on its ends are especially large. When this finally occurred, sieve tube members came into being (Fig. 7.13).

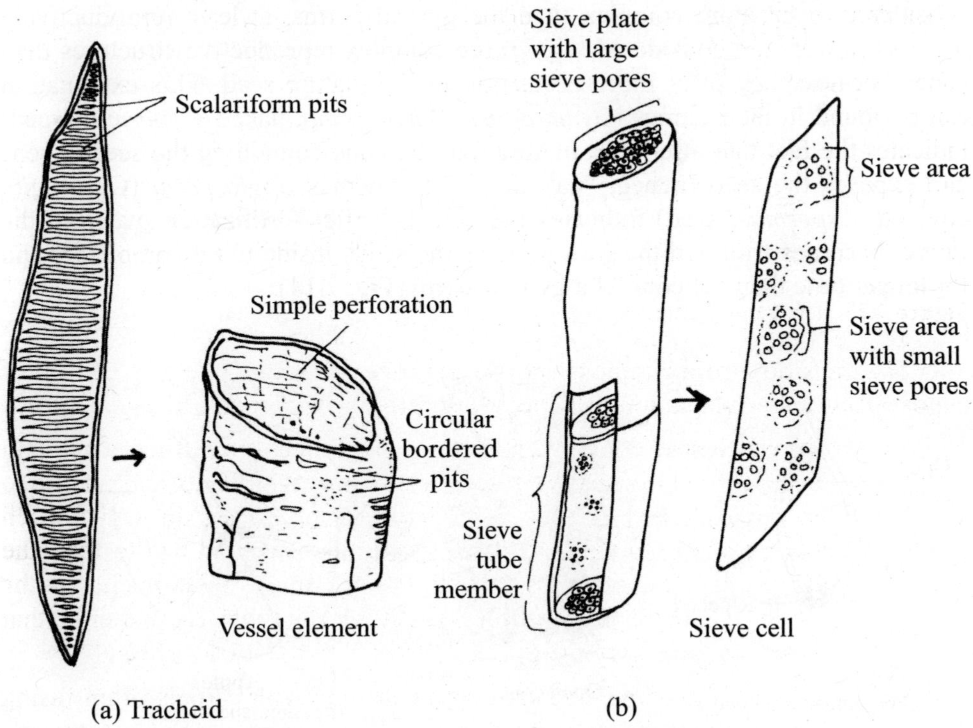

Figure 7.13. (a) Tracheid and vessel element (b) Sieve tube members

7. Leaves became broader and developed reticulate venation, and they became more polymorphic, adaptable to a variety of functions, not just photosynthesis.
8. The most fundamental transition from gymnosperm to angiosperm was the acquisition of developmental plasticity in angiosperms, i.e. the capacity to survive mutations that alter growth and development. Gymnosperms generally lack developmental plasticity, for example, conifers. They all are large, giant trees with one or few main stems and needle- or scale-shaped leaves and greatly resembles each other. This uniformity is even more striking when one considers that gymnosperms dominated earth for hundreds of millions years and could have diversified into almost any habitat. On the other hand, angiosperms shows tremendous developmental plasticity, beginning about 125 to 145 million years ago. They have diversified so greatly that not only hundreds of thousands of species, but also thousands of types, now exist. They range from gigantic to tiny, woody to herbaceous, perennial to biennial to annual, temperate to tropical, desert habitats to mesic to rainforest to aquatic, autotrophic to partially parasitic to fully parasitic, and epiphytic to subterranean to endophytic.

At first glance, the cone-bearing gymnosperms would seem to be the equal of the flower-forming angiosperms, but such is not the case. The angiosperms are considered to be more complex than the gymnosperms, at least reproductively because flowers are considered to be more complex reproductive structures than cones because they offer more protection to the mature seed. This explanation can be found in their names: *Gymnosperm* (Gr. *Gymno*, naked + *spermae*, seed) indicates the fact that after seed maturation, the cone containing the seeds opens and exposes the seed (hence, "naked seed"), whereas *Angiosperm* (Gr. *Angio*, covered + *spermae*, seed) indicates the fact that after fertilization ovary of the flower becomes fruit and the fruit protects the seeds inside more completely and for longer time than the cone of a gymnosperm (Fig. 7.14).

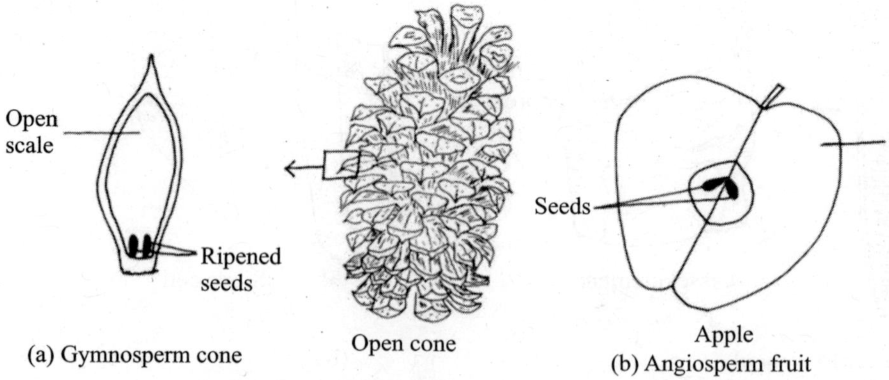

Figure 7.14. (a) Gymnosperm naked seed (b) Angiosperm covered seeds

The evolutionary changes involved in the conversion of a gymnosperm line into an angiosperm line did not occur instantaneously, nor did they occur in all species. The ability to form flowers and double fertilization were probably the first transformation, because these features are unique to all angiosperms. Sieve tubes evolution in flowering plants was possibly the next transformation because only one or few species are thought to lack them. But the flowering plants must have existed and diversified for several million years before vessel elements in xylem evolved, because several families of flowering plants lack this feature; these families must have evolved and become distinct before vessels appeared. Today vessels are present in all herbaceous species, so herbaceousness must have developed next, and it probably arose several times in separate lines. The ability to form flowers in the first year after seed germination is a necessary trait for annual plants, most of which are herbs. Characteristics like succulence, parasitism evolve much later, as very few isolated families or species have these characters. Other derived features are the fusion of the petals into one structure (sympetally) and floral zygomorphy, i.e. flowers are bilaterally symmetrical not radially symmetrical. Some of the relictual and derived features in flowering plants are given below in Table 7.1.

Table 7.1. Relictual and derived features in Division Anthophyta

	Relictual features (present in early angiosperms)	**Derived features (present only in some later, more modern angiosperms)**
Habit		
Size	Large bush, small tree	Large trees, herbs, bulbs, vines, many types
Leaf retention	evergreen	Deciduous or leafless
Wood		
Vessels	None	Present
Axial parenchyma	None or little	Abundant
Rays	All narrow, tall	Of several types in one plant: short/wide and narrow/tall
Flowers		
Presence of parts	Complete flowers	Incomplete/imperfect
Number of each type of part	Many	Few: sets of 3,4,5
Arrangement	Spirals	Whorls
Symmetry	Radial	Bilateral
Fusion of parts	None	Much fusion
Position of ovary	Superior	Inferior
Pollination	Wind/beetles	Many types
Seed dispersal	wind	Many types

Flowering plants had diversified so much in the past that it should not be thought that the evolution in these group of plants comes to an end. Some groups do seem to be changing very little now, but others, especially grasses, orchids and composites are still changing and evolving so rapidly that it is difficult to keep up with them. Despite their diversity and current evolutionary process, angiosperms are clearly united by a suite of synapomorphies (i.e., shared derived features) including:

1. Ovules that are enclosed within a carpel, i.e. a structure that is made of an ovary that encloses the ovules, and the stigma, a structure where pollen germination take place.
2. Double fertilization, which leads to the formation of an endosperm.
3. Stamens with two pairs of pollen sacs.
4. Features of gametophyte structure and development
5. Phloem tissue composed of sieve tubes and companion cells.

7.5 ANGIOSPERM PHYLOGENY

Traditionally, the angiosperms have been classified into two groups on the basis of the number of cotyledons in their seeds – monocots (one) and dicots (two). However recent molecular studies have revealed that dicots are not monophyletic but represent several lineages instead. Based on molecular data, the phylum Anthophyta contains eight clades: *Amborella*, *Nympheaceae*, *Austrobaileyales*, *Chloranthaceae*, *Magnoliids*, *Monocotyledons*, *Ceratophyllaceae*, and *Eudicots* (Fig. 7.15).

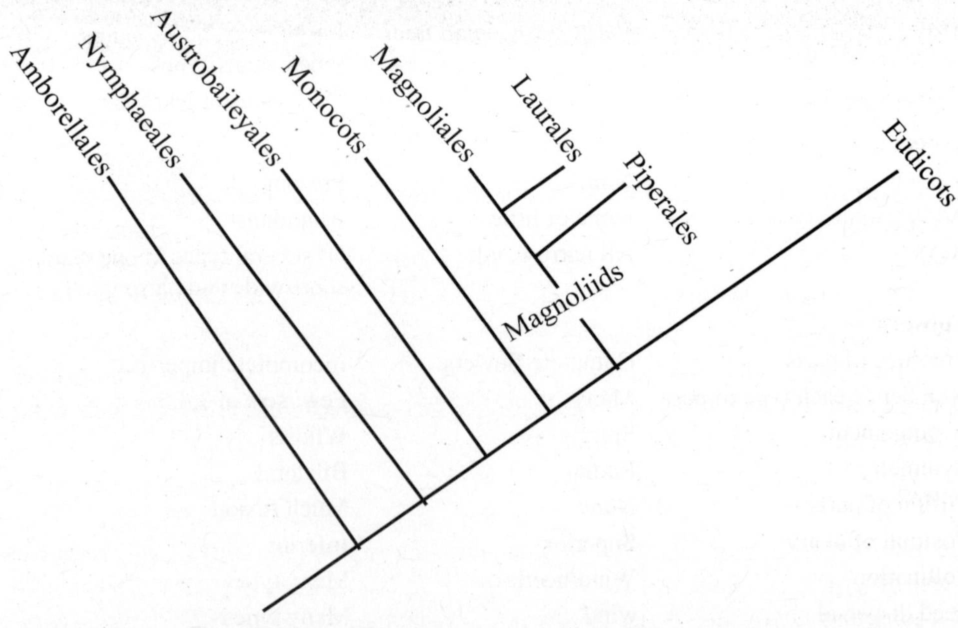

Figure 7.15. Possible evolutionary relationships of major groups of Angiosperms in the form of Cladogram

7.5.1 The basal angiosperms

The basal angiosperms represent a grade that includes the following groups: Amborellaceae, Nymphaeaceae, Austrobaileyales, Ceratophyllaceae, Chloranthaceae, magnoliids, and monocots (although not all investigators choose to consider monocots basal angiosperms).

The exact relationship between these eight groups is not yet clear, although there is general agreement that the first three groups to diverge from the ancestral angiosperm were ***Amborellaceae***, ***Nymphaeaceae***, and ***Austrobaileyales***. As such, the term **basal angiosperms** is applied to these three groups. Basal angiosperms, however, are not considered a monophyletic grouping.

The basal angiosperms are not newly discovered species in that they had been classified as monocots or eudicots before molecular analysis revealed they had become reproductively separate from the rest of angiosperms very early during angiosperm evolution. The evolutionary relationships of these groups are discussed below.

Amborella: The root of Angiosperm evolutionary tree

A mere decade ago, the possibility of identifying the basal nodes of the angiosperm clade seemed remote. However, most analyses of the past five years concur in placing the monotypic *Amborella* as the sister to all other extant angiosperms, *Amborella trichopoda* (Fig. 7.16), endemic to cloud forests of New Caledonia, was described in the mid-19th century and has since been classified with various groups of basal angiosperms, most often with Laurales, a group of magnoliids. However, *Amborella* lacks those features considered to be synapomorphies for Laurales and clearly differs from most Laurales in having spirally arranged floral organs, rather than the whorled arrangement typical of most Laurales. *Amborella* has carpels that are closed only by secretion, rather than by fused tissue as in most angiosperms–a feature that may represent a plesiomorphy (i.e., ancestral feature) for the angiosperms. Features that unify all extant angiosperms except *Amborella* include vessels and pollen grains with a reticulate tectum. Synapomorphies for all extant angiosperms except *Amborella* and Nymphaeaceae include ethereal oil cells–common throughout basal angiosperms–and columellate pollen grains with a perforate tectum.

The placement of *Amborella* as sister to all other angiosperms is supported by nearly all multigene analyses of basal angiosperms, including evidence from all three plant genomes.

Nymphaeaceae

The phylogenetic position of Nymphaeaceae as one of the two basal-most (or oldest diverging) lineages of extant angiosperms is strongly supported by nearly all molecular analyses. This clade of eight aquatic genera has a worldwide distribution, consistent with the ancient age of this lineage based on the fossil record (Fig. 7.17).

Figure 7.16. *Amborella trichopoda* (Amborellaceae)

Figure 7.17. *Nuphar japonica* (Nymphaeaceae)

Austrobaileyales

The positions of Amborellaceae and Nymphaeaceae as successive sisters to the rest of the angiosperms are followed, in turn, by Austrobaileyales. Austrobaileyales are a small clade that comprises Austrobaileyaceae (*Austrobaileya*) and Trimeniaceae (*Trimenia*). Although the traditional Illiciaceae and Schisandraceae have typically been considered to be closely related, a relationship between these taxa and *Austrobaileya* and *Trimenia* had not been suspected. No morphological characteristics have been identified that unify this group, despite the strong molecular support for its monophyly (Fig. 7.18).

Figure 7.18. *Austrobaileya* sp. (Austrobaileyales)

Relationships among Ceratophyllaceae, monocots, Chloranthaceae, magnoliids, and eudicots

Resolution and support for relationships among Ceratophyllaceae, monocots, Chloranthaceae, magnoliids, and eudicots are poor. Individually, each lineage is well supported, and both the fossil record and molecular-based trees identify these lineages as ancient. However, relationships among them remain unclear. It is clear, however, that angiosperms do not fall into two major groups that correspond to monocots (Liliopsida) and dicots (Magnoliopsida) of most traditional classification systems such as Cronquist (1981), Takhtajan (1997), and their predecessors. Although monocots clearly form a strongly supported group, dicots in the traditional sense do not: most are found in the eudicot clade, but the remaining nonmonocot basal branches (i.e., Amborellaceae, Nymphaeaceae, Austrobaileyales, Ceratophyllaceae, Chloranthaceae, magnoliids) were also "traditional" dicots. It was long suspected that the "dicots" as traditionally recognized represents multiple evolutionary lineages, and the nonmonophyly of dicots based on molecular data precludes their recognition in current classifications. Thus, the concept of "dicot" should be abandoned in favor of eudicots.

Monocots

Among extant angiosperms, monocotyledons represent the earliest-appearing major clade. Molecular clock suggest that the origin of the monocot clade to be 134 mya older than the oldest angiosperm fossils. Although their exact age is unclear from the fossil record, monocots clearly represent an early lineage of angiosperms. There are approximately 52000 species of monocots, representing 22% of all angiosperms. Half of the monocots can be found in the two largest families, Orchidaceae and Poaceae, which comprise 34% and 17%, respectively, of all monocots

While dicots are not a valid group, the monocots indeed are a distinct group within the angiosperms. Ray (1703) first identified the monocots as a group, based largely on their possession of a single cotyledon. Nonmolecular phylogenetic studies of monocots support this grouping; these studies have identified 13 putative synapomorphies for the monocots, including, among others, a single cotyledon, parallel-veined leaves, sieve cell plastids with several cuneate protein crystals, scattered vascular bundles in the stem, and an adventitious root system. An often-overlooked synapomorphy for monocots is their sympodial growth; although there are other angiosperms with sympodial growth, monocots are nearly exclusively so.

Another major distinctive trait of the monocots is their vascular system, which is characterized by vascular bundles that are scattered throughout the medulla and cortex and are closed (i.e., do not contain an active cambium). In contrast, basal angiosperms formerly considered dicots (e.g., members of the magnoliid clade) and eudicots possess open vascular bundles arranged in a ring.

In some analyses the monocots appear as the sister to Ceratophyllaceae, with the clade of monocots + Ceratophyllaceae sister to Chloranthaceae + magnoliids + eudicots. However, relationships of monocots are unclear and further work is needed to assess the validity of their relationships.

Ceratophyllaceae

Ceratophyllaceae (*Ceratophyllum*) had the distinction of appearing as the sister to all other angiosperms in the first large molecular phylogenetic analysis based on *rbcL* (Chase *et al.*, 1993). The aquatic habit and simple flowers seemed at odds with most hypotheses about the earliest angiosperms, although *Ceratophyllum* has a long fossil record, going back at least 125 million years ago. Subsequent analyses demonstrated that this placement was unique to the *rbcL* data set. Our current understanding of the relationship of *Ceratophyllum* to other angiosperms, based on evidence from many other genes, is still not clear although it may be related to the monocots (Fig. 7.19).

Figure 7.19. *Ceratophyllum demersum* (Ceratophyllaceae)

Chloranthaceae

Chloranthaceae, with their small, simple flowers, have an extensive fossil record, dating back 125 million years. However, although the origins of the family are ancient, most extant genera are relatively young in age. Chloranthaceae are clearly an isolated lineage separate from the magnoliid clade, but their phylogenetic position remains uncertain. In some analyses, they are sister to a clade of magnoliids + eudicots, albeit with weak support (Fig. 7.20).

Figure 7.20. *Sarcandra chloranthoides* (Chloranthaceae)

Figure 7.21. *Magnolia grandiflora* (Magnoliaceae)

Magnoliids

The magnoliid clade comprises most of those lineages typically referred to as "primitive angiosperms" in earlier works (e.g., Cronquist, 1981, 1988; Stebbins, 1974; Takhtajan, 1997). Although the component families of the magnoliid clade were loosely associated in previous classifications, for example, as Cronquist's

(1981), subclass Magnoliidae, relationships among the families were not clear. In addition, Magnoliidae contained groups that are not part of the magnoliid clade as recognized by phylogenetic analyses. Reconstructing relationships within this clade is challenging, given the age of the group (some putative members, such as *Archaeanthus*, Dilcher and Crane, 1984, date to the early Cretaceous) and presumably high levels of extinction. Although the major lineages of the magnoliid clade were identified as well-supported clades in earlier studies. Some phylogenetic analyses weakly support a grouping of magnoliids as sister to the eudicots, although more data are necessary to clarify this relationship. Within the magnoliids, Magnoliales (Fig. 7.21) and Laurales are sisters, and Piperales and Canellales are sisters.

Eudicots

Eudicots, a clade strongly supported by molecular data, comprise the bulk of angiosperm species (approximately 75%). This large clade is supported by a single morphological synapomorphy–triaperturate pollen. This pollen type is unique to the eudicots, and while not all eudicots have triaperturate pollen due to subsequent changes in pollen structure, triaperturate pollen is clearly distinct from the uniaperturate pollen of basal angiosperms, monocots, and all other seed plants, allowing easy assignment of fossil pollen to the eudicots. The fossil pollen record indicates that the eudicots appeared 125 million years ago, shortly after the origin of the angiosperms themselves. The extensive fossil pollen collections worldwide, coupled with solid dates, make it unlikely that the eudicots arose much before this time point.

A basal grade of five lineages (Ranunculales, Proteales, Sabiaceae, Trochodendraceae, and Buxaceae) subtends the large clade of core eudicots. Although Ranunculales are supported as the sister to all other eudicots, the relative placements of the remaining four lineages of basal eudicots are not clear and require additional study. The core eudicots comprise the vast majority of eudicot species. Seven major clades (Gunnerales, Berberidopsidales, Saxifragales, Santalales, Caryophyllales, Rosids and Asterids) have been recognized.

The super tree approach

A recent super tree analysis combined trees that included all angiosperm families and produced the first comprehensive family-level phylogenetic tree for angiosperms. The basic framework of the angiosperm super tree is largely consistent with the results of large, multigene analyses of exemplar taxa on which it was based. *Amborella* is sister to all other angiosperms, followed by Nymphaeaceae, Austrabaileyales, a clade of monocots + Ceratophyllaceae, Chloranthaceae, and a clade of magnoliids + eudicots.

SUBJECT INDEX

Abiogenesis 7
ABO blood groups 119
Absolute dating 42
Acanthodians 164
Adaptation 24, 88
Adaptive divergence 114
Adaptive evolution 124
Adaptive radiation 32, 33, 81, 107, 170
Aegyptopithecus 189
Agnatha 162
AIDS 30
Algae 4
Allantois 172
Allele frequency 65
Alleles 55
Allopatric speciation 105
Allopolyploid 111
Allozymes 119
Alvarez Theory 85
Amber 38
Amborella 216, 217
Amino acids 8
Amnion 172
Amniotic egg 162, 172
Amniotic fluid 172
Amphibia 162
Amphibians 169
Anagenesis 96
Analogies 31
Analogous structures 46
Anatomically modern humans 195
Anatomy 45
Ancestral traits 132
Angiosperms 202
Ankylosaurus 210
Anthophyta 202
Anthropoidea 184
Antibodies 119
Anura 170
Apatosaurus 210
Apoda 170
Archaeopteryx 40
Archaeopteryx lithographica 179
Archaeothyris 79
Archipelagos 33
Archoeozoic 43
Ardipithecus ramidus 191
Artificial selection 26
Artiodactyla 183
Atmosphere 8
ATP 12, 49
Australopithecus afarensis 191
Australopithecus africanus 191
Australopithecus anamensis 191
Australopithecus robustus 194
Austrobaileya 211
Austrobaileyales 216, 218
Autopolyploid 111
Aves 162
Back cross 56
Bacteria 2
Balanced polymorphism 121
Basal angiosperms 217
Behavioural isolation 102
Biogenesis 6
Biogenetic law 48
Biogeography 31
Biological species concept 97

Bipedality 190
Birds 7
Bottleneck effect 72
Brachiosaurus 210
Bryophytes 201
Bryophytes 6
Cambrian explosion 83
Cambrian period 5
Carinates 180
Carnivora 183
Caytonia 207
Cenozoic 43
Cephalochordata 159
Ceratophyllaceae 216, 220
Cetacea 183
Character 53
Charophyceae 201
Charophyceans 6
Chelonia 176
Chemical evolution 17
Chemosynthetic origin 7
Chiroptera 183
Chloranthaceae 216, 220
Chloroplast 4
Chondrichthyes 162
Chordata 156
Chorion 172
Chromosomal mutation 29
Clade 141
Cladogenesis 88, 97
Clarksonia 201
Cline 109
Cnidarians 5
Coacervate 13
Coadaptation 90
Codominance 60
Coefficient of coincidence 60
Coefficient of Interference 60
Coevolution 86
Comparative embryology 31
Competitive exclusion 34
Continental drift 9, 34, 37
Convectional currents 9
Convergent evolution 32, 47, 134
Copy number polymorphisms 119
Core 9
cpDNA 132
Cretaceous extinction 84, 85
Crocodilia 176
Crust 9
Cyanobacteria 2
Cycadeoids 208
Cynodonts 80
Cytochrome c 130
Cytoplasmic inheritance 63
Darwin 20, 23
Darwins finches 27
DDT 30
Deep-sea vents 5, 17
Degenaria 211
Deme 64
Derived trait 134
Developmental plasticity 214
Differentiation 1, 9
Dihybrid cross 56
Dimetrodon 80
Dinosaur 38
Dinosaurs 174
Diplodocus 210
Directional selection 93
Disassortative mating 76
Disruptive selection 95
Divergent evolution 46
Diversifying selection 93
DNA 14, 49
Dominant 54
Double fertilization 212
Earth 1
Ecological niche 31, 89
Ecological species concept 100
Ecotone 109
Ectotherms 174
Electrophoresis 119
Embryology 48

Endosymbiotic theory 4
Endothermic 177
Enzymes 12
Eohippus 40
Eucalyptus 203
Eudicots 216, 222
Eukaryotes 4
Europa 17, 19
Eutherian 181
Evolution 18
Evolutionary reversal 134
Exons 125
External fertilization 103
Extinction 80
Extinction rates 33
Fishes 6
Foregut fermentation 146
Fossil 3
Fossils 31, 37
Founder effect 73, 120
Fungi 6
Galapagos finches 107
Galapagos islands 33
Gametic isolation 103
Gause's competitive exclusion principle 100
Gene duplication 145
Gene family 150
Gene flow 69, 74
Gene mutation 29
Gene pool 29, 64
Gene tree 151
Genes 28, 55
Genetic change 90
Genetic code 50
Genetic drift 28, 29, 69, 71, 120
Genetic recombination 29
Genetic structure 65
Genetics 52
Genome 145
Genotype 14, 55
Genotypic frequency 65
Genus 140
Geographical barrier 101
Geological time 3
Geological time-scale 43
Gizzard 177
Gnathostomata 162
Gorilla 186
Greenland 1
Gymnosperms 202
HAART 153
Habitat 47
Habitat isolation 102
Habitat tracking 89
Haldane 7
Hardy-Weinberg equilibrium 67
Hardy-Weinberg theorem 65
Hemoglobin 50, 151
Heredity 14
Heritable 53
Heterozygous 55
Himantandra 211
Hitch-hiking 115
HIV 30
HMS Beagle 23
Homeostasis 89
Homo erectus 194
Homo habilis 193
Homo sapiens 184, 189, 195
Homologies 31
Homologous structures 46
Homology 46
Homoplasies 134
Homoplastic traits 135
Homozygous 55
Host shift 114
Hot springs 5
Hox genes 163
Human major 119
Human major histocompatibility complex 119
Huxley 28
Hybrid breakdown 104

Hybrid inviability 103
Hybrid sterility 103
Hybrid zone 110
Hydrothermal vents 17
Hylobates 186
Hymenopterans 209
Hyoid 164
Hyracotherium 40
Ichthyostega 40
Immigration rates 33
Inbreeding 75
Inbreeding depression 75
Incomplete dominance 60
Independent assortment 57
Index fossils 42
Industrial melanism 94
Ingroup 135
Insecticide resistance 30
Insectivora 183
Interference 60
Internal fertilization 103
Intersexual selection 26
Interspecific 25
Intrasexual selection 26
Intraspecific 25
Introns 125
Invertebrate chordates 157
Invertebrates 149
Iridium 85
Island biogeography 33
Isozyme 124
Kimura 124
Koonwarra angiosperm 204
Lagomorpha 183
Lamarck 20, 21
Lepidopterans 209
Linkage 58, 59
Lysozyme 146
Macroevolution 28, 77
Magma 10
Magnolia 205, 211
Magnoliaephyllum 204
Magnoliids 216, 221
Mammalia 162
Mammals 6
Mantle 9
Marsupials 181
Mass extinctions 83
Mechanical isolation 103
Megazostrodon 79
Membrane potential 14
Mendel 52
Mesozoic 43
Mesozoic era 37
Metabolism 14
Metalloenzymes 13
Metamorphosis 171
Meteorite 8
Meteorites 17
Micro-evolution 28, 70
Microsphere 13
Miller 11
Missing link 40
Mitochondria 4
Modern synthesis 28, 64
Molecular clock 4, 125, 127, 131
Molecular evolution 14, 124, 125
Molecular traits 136
Monocots 219
Monocotyledons 216
Monohybrid cross 54
Monophyletic 141
Monotremes 181
Montanoceratops 210
Morphology 135
mtDNA 132, 197
Multiple alleles 28, 61
Mutations 29, 74, 123
Myoglobin 151
Natural selection 14, 15, 24, 25, 29, 76
Naturalistic Theory 8
Neo-Darwinism 28
Nerve Cord 156
Neural crest 160

Neural tube 160
Neutral genes 123
Neutral mutants 124
New World monkeys 185
Nitrogenous bases 11
Non-random mating 75
Non-synonymous mutation 127
Notochord 135, 156
Nucleic acids 8
Nucleotides 8
Nymphaeaceae 216, 217
Old World monkeys 185
Oleanane 203
Ontogeny 48
Oparin 7
Organic monomers 11
Organic polymers 12
Orthogenesis 38, 39
Orthologs 152
Osteichthyes 162
Outcrossing 144
Outgroup 135
Oviparous 166
Ovoviviparous 167
Oxygen revolution 4
Ozone 8
Paedogenesis 171
Paleontology 31
Paleozoic 43
Paleozoic era 5
Pan 186
Pangea 35
Pangenesis 25
Pangenesis 27
Panselectionism 123
Panspermia 17
Paralogs 152
Parapatric speciation 105, 109
Paraphyletic 141
Parasaurolophus 210
Parsimony principle 138, 139
Particulate theory 55
Pedigrees 59
Pelycosaurs 79
Peripheral isolate 106, 117
Perissodactyla 183
Permian extinction 84
Pharyngeal Slits 156
Phenetic species concept 99
Phenotype 14, 55
Phenotypic assortative mating 76
Phyletic gradualism 116
Phylogenetic tree 132
Phylogeny 39, 48, 132
Placoderms 164
Planetesmials 8
Pleiotropy 62, 115
Polygenic inheritance 61
Polygenic traits 61
Polymorphism 119
Polyphyletic 141
Polyploidy 111
Pongo 186
Population 27, 64
Population genetics 64
Poriferans 5
Postnatal Tail 157
Precambrian era 4
Precambrian period 5
Primates 184
Prion proteins 121
Probainognathus 80
Proboscidea 183
Prokaryotic 1
Prophage 30
Prosimii 184
Proteins 8
Protenoids 13
Proterozoic 43, 45
Protobionts 8
Pseudogene 150
Pseudogenes 128, 149
Pteridophytes 6, 201
Pterosaurs 174

Punctuated equilibrium 116
Punnett square 55
Quantitative characters 64
Quantitative traits 61
Radiocarbon Dating 42
Radiometric dating 42
Random mating 69
Ratites 180
rDNA 132
Recessive 54
Recognition species concept 99
Red Queen hypothesis 90
Reducing atmosphere 16
Relative dating 42
Reproductive isolation 29, 101
Reptiles 6
Reptilia 162
Restriction Fragment Length Polymorphisms 119
Retrovirus 30
Reverse transcriptase 30
Ribozyme 14
RNA 3, 14
Rodentia 183
rRNA 144
Sauropsids 174
Secondary sexual characteristics 26
Sedimentary rocks 37
Selectionism 123
Self-compatible 144
Self-fertilization 52
Self-incompatibility 144
Selfing 54
Sex chromosomes 63
Sexual dimorphism 26, 190
Sexual selection 25, 26
Shared derived traits 138
Shared mate recognition system 99
Sickle-cell anemia 93
Sidney Fox 13
Sieve tubes 213
Silent mutation 126
Similarly matrix 128
Single Nucleotide Polymorphisms 119
Sirenia 183
Snowball earth 5
Solar system 16
Speciation 28, 77, 96
Species 20, 25, 64
Species equilibrium model 33
Species selection 118
Sporophylls 211
Squamata 176
Stabilizing selection 93
Stepped clines 109
Stromatolites 1
Subduction zone 35
Sympatric speciation 105, 111
Synapomorphies 216
Synapsida 79
Synapsids 174
Taxonomy 31
Temporal fenestra 80
Temporal isolation 102
Tension zone 110
Test cross 56
Tetrapods 6
Therapsids 80
Tissue typing 120
Tracheids 213
Trait 53
Transformism 21
Transposable elements 149
Triticum aestivum 114
True breeding 53
Ultraviolet radiation 8
Universe 17
Urey 11
Urochordata 157
Urodela 170
Variatins 25
Variations 29
Vents 10
Vertebral column 160

Vertebrates 149, 159
Vessel elements 213
Vestigial organs 47
Viviparous 167
Wallace 24
Wolffia 203
Yolk sac 172